Introduction to Infinite-Equilibriums in Dynamical Systems

Albert C. J. Luo

Introduction to Infinite-Equilibriums in Dynamical Systems

Springer

Albert C. J. Luo
Department of Mechanical and Mechatronics Engineering
Southern Illinois University Edwardsville
Edwardsville, IL, USA

ISBN 978-3-031-89085-7 ISBN 978-3-031-89083-3 (eBook)
https://doi.org/10.1007/978-3-031-89083-3

This Springer imprint is published by the registered company Springer Nature Switzerland AG
The registered company address is: Gewerbestrasse 11, 6330 Cham, Switzerland

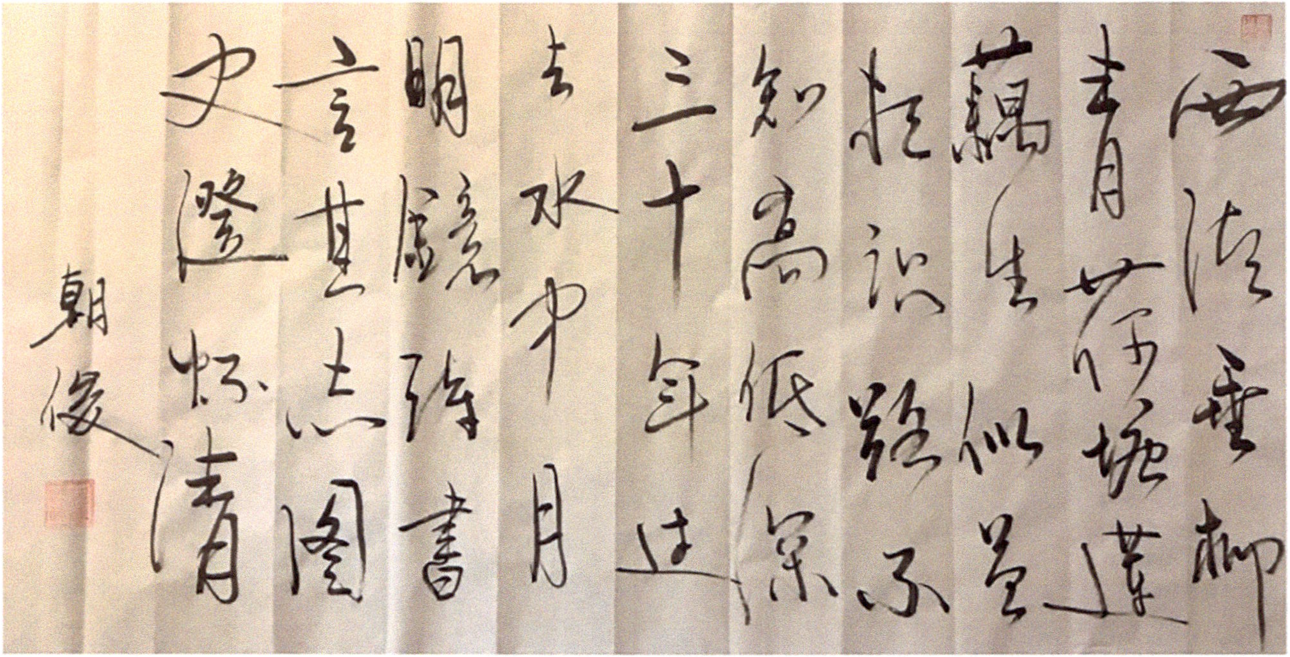

西湖垂柳青，荷塘莲藕生，似曾相识路，不知高低深。三十年过去，水中月明镜，诗书言其志，图史澄怀清。 — 朝俊于癸卯《夜探唐宝住院》

Preface

This book is about infinite equilibriums for the switching bifurcations of two 1-dimensional flows in dynamical systems. The infinite equilibriums in dynamical systems were initially discussed by the author in his work in 2021, "Two-Dimensional Quadratic Nonlinear Systems, Volume II: Bivariate Vector Fields," which was published in 2022. However, how to determine the infinite equilibriums with linear-bivariate vector fields was not clear, and the parabola source and sink infinite equilibriums were mistakenly defined as the parabola-saddle infinite equilibriums. In this book, the quadratic systems with a linear-bivariate function are also adopted to discuss infinite equilibriums in dynamical systems. For such quadratic dynamical systems, there are three types of infinite equilibriums. The inflection-source and sink infinite equilibriums are for the switching bifurcations of two parabola flows on the two directions. The parabola-source and sink infinite equilibriums are for the switching bifurcations of parabola and inflection flows on the two directions. The inflection upper- and lower-saddle infinite equilibriums are for the switching bifurcation of two inflection flows in two directions. The inflection flows are for appearing bifurcations of two parabola flows on the same direction. Such switching bifurcations for 1-dimensional flow are based on the infinite equilibriums, which will help one understand global dynamics in nonlinear dynamical systems. This book will introduce infinite equilibrium concepts and such switching bifurcations to nonlinear dynamics.

In Chap. 1, single-linear-bivariate linear systems are discussed first, and the switching bifurcations for the 1-dimensional flows (e.g., sink and source flows, and up-parabola and down-parabola flows) are presented. The increasing-inflection and decreasing-inflection sink and source infinite equilibriums are for the switching bifurcations of the up-parabola and down-parabola flows.

In Chap. 2, a quadratic dynamical system possessing constant and linear-bivariate quadratic vector fields is discussed, and the corresponding inflection flows exist for the appearing bifurcation of up- and down-parabola flows.

In Chap. 3, dynamical systems with single-linear-bivariate linear and quadratic vector fields are considered. Parabola-source and sink infinite equilibriums for the switching of inflection and parabola flows are discussed, and the inflection-source and sink infinite equilibriums for parabola flows in two directions are presented. The inflection flows are for the appearing bifurcation of up- and down-parabola flows. The 1-dimensional parabola flow arrays exist.

In Chap. 4, dynamical systems possessing two quadratic vector fields with a single-linear-bivariate function are discussed. Two inflection flows exist in such quadratic systems for appearing bifurcations of up- and down-parabola flows. The parabola-source and sink infinite equilibriums are for the switching bifurcations of parabola and inflection flows. The inflection-source and sink infinite equilibriums are for the switching bifurcations of two parabola flows in two directions. The inflection upper- and lower-saddle infinite equilibriums are for the switching bifurcations between two inflection flows in two directions. The corresponding parabola flow arrays exist in such quadratic systems.

Finally, the author hopes that the materials presented herein can provide a better understanding of switching bifurcations of two 1-dimensional flows through the infinite equilibriums in nonlinear dynamical systems.

Edwardsville, IL, USA Albert C. J. Luo

Contents

Chapter 1
Single-Linear-Bivariate Linear Systems

In this chapter, single-linear-bivariate linear systems will be discussed, and the switching bifurcations for the one-dimensional flows (e.g., sink and source flows and up-parabola and down-parabola flows) will be presented. The increasing-inflection and deceasing-inflection sink and source infinite-equilibriums are for the switching bifurcations of the up-parabola and down-parabola flows.

1.1 Infinite-Equilibriums and Parabola Flows

In this section, infinite-equilibriums and one-dimensional flow in linear-bivariate linear dynamical systems will be discussed. The switching of two one-dimensional parabola flows is discussed through infinite-equilibriums.

Theorem 1.1 *Consider a two-dimensional, bivariate linear system as*

$$\begin{aligned} \dot{x}_{j_2} &= A_{j_2 j_2 1} + A_{j_2 j_2 0} y_{j_1}, \\ \dot{x}_{j_1} &= A_{j_1 j_1 1} + A_{j_1 j_1 0} y_{j_1}. \end{aligned} \tag{1.1}$$

with

$$\begin{aligned} & y_{j_1} = \alpha_{j_1 j_1} x_{j_1} + \alpha_{j_1 j_2} x_{j_2} \text{ for } j_1, j_2 \in \{1,2\}, j_1 \neq j_2; \\ & \mathbf{n}_{j_1} = (\alpha_{j_1 j_1}, \alpha_{j_1 j_2})^{\mathrm{T}}. \end{aligned} \tag{1.2}$$

(i) If $A_{j_2 j_2 1} = A_{j_2 j_2 0} = 0$, the two-dimensional dynamical system becomes a one-dimensional linear bivariate system in the x_{j_1}-direction. The one-dimensional system becomes

A. C. J. Luo, *Introduction to Infinite-Equilibriums in Dynamical Systems*, https://doi.org/10.1007/978-3-031-89083-3_1

$$\begin{aligned} \dot{x}_{j_2} &= 0, \\ \dot{x}_{j_1} &= a_{j_1j_10}(y_{j_1} - a_{j_1j_11}). \end{aligned} \tag{1.3}$$

with

$$a_{j_1j_11} = -\frac{A_{j_1j_11}}{A_{j_1j_10}}, a_{j_1j_10} = A_{j_1j_10}. \tag{1.4}$$

There is a dynamical system in the y_{j_1}-direction *as*

$$\dot{y}_{j_1} = \alpha_{j_1j_1}a_{j_1j_10}(y_{j_1} - a_{j_1j_11}), \tag{1.5}$$

with a dynamical system in the $y_{j_1}^{\perp}$*-direction as*

$$\dot{y}_{j_1}^{\perp} = \beta_{j_1j_1}a_{j_1j_10}(y_{j_1} - a_{j_1j_11}), \tag{1.6}$$

where

$$\begin{aligned} &y_{j_1}^{\perp} = \beta_{j_1j_1}x_{j_1} + \beta_{j_1j_2}x_{j_2}, \mathbf{n}_{j_1}^{\perp} = (\beta_{j_1j_1}, \beta_{j_1j_2})^{\mathrm{T}}; \\ &\mathbf{n}_{j_1} \cdot \mathbf{n}_{j_1}^{\perp} = \alpha_{j_1j_1}\beta_{j_1j_1} + \alpha_{j_1j_2}\beta_{j_1j_2} = 0. \end{aligned} \tag{1.7}$$

Thus, the flow along the y_{j_1} *-direction is*

$$t - t_0 = \frac{1}{\alpha_{j_1j_1}a_{j_1j_10}} \ln \frac{|y_{j_1} - a_{j_1j_11}|}{|y_{j_10} - a_{j_1j_11}|}. \tag{1.8}$$

- *For* $a_{j_1j_10}\alpha_{j_1j_1} > 0$, the line of $y_{j_1}^* = a_{j_1j_11}$ in the x_{j_1} *-direction is a source infinite-equilibrium.*
- *If* $a_{j_1j_10}\alpha_{j_1j_1} < 0$, the line of $y_{j_1}^* = a_{j_1j_11}$ in the x_{j_1} *-direction is a sink infinite-equilibrium.*

(ii) If

$$A_{j_2j_21} \neq 0, A_{j_2j_20} = 0 \tag{1.9}$$

the standard form is

$$\begin{aligned} \dot{x}_{j_2} &= A_{j_2j_21}, \\ \dot{x}_{j_1} &= a_{j_1j_10}(y_{j_1} - a_{j_1j_11}). \end{aligned} \tag{1.10}$$

There is a dynamical system in the y_{j_1}-direction *as*

$$\dot{y}_{j_1} = \alpha_{j_1 j_2} A_{j_2 j_2 1} + \alpha_{j_1 j_1} a_{j_1 j_1 0}(y_{j_1} - a_{j_1 j_1 1}), \tag{1.11}$$

with a dynamical system in the $y_{j_1}^{\perp}$*-direction as*

$$\dot{y}_{j_1}^{\perp} = \beta_{j_1 j_2} A_{j_2 j_2 1} + \beta_{j_1 j_1} a_{j_1 j_1 0}(y_{j_1} - a_{j_1 j_1 1}), \tag{1.12}$$

where

$$\begin{aligned} & y_{j_1}^{\perp} = \beta_{j_1 j_1} x_{j_1} + \beta_{j_1 j_2} x_{j_2}, \mathbf{n}_{j_1}^{\perp} = (\beta_{j_1 j_1}, \beta_{j_1 j_2})^{\mathrm{T}}; \\ & \mathbf{n}_{j_1} \cdot \mathbf{n}_{j_1}^{\perp} = \alpha_{j_1 j_1} \beta_{j_1 j_1} + \alpha_{j_1 j_2} \beta_{j_1 j_2} = 0. \end{aligned} \tag{1.13}$$

(ii$_1$) Let

$$\frac{\alpha_{j_1 j_2} A_{j_2 j_2 1}}{\alpha_{j_1 j_1} a_{j_1 j_1 0}} \equiv b_{j_1 j_1 1}, \dot{y}_{j_1}^{\perp *} = \frac{A_{j_2 j_2 1}}{\alpha_{j_1 j_1}} (\alpha_{j_1 j_2}^2 + \alpha_{j_1 j_2}^2) \tag{1.14}$$

then the flow along the y_{j_1} *-direction is*

$$t - t_0 = \frac{1}{\alpha_{j_1 j_1} a_{j_1 j_1 0}} \ln \frac{| y_{j_1} - a_{j_1 j_1 1} + b_{j_1 j_1 1} |}{| y_{j_1 0} - a_{j_1 j_1 1} + b_{j_1 j_1 1} |}, \tag{1.15}$$

with the equilibriums of $y_{j_1}^* - a_{j_1 j_1 1} = -b_{j_1 j_1 1}$ in the y_{j_1}-direction.

- If $\alpha_{j_1 j_1} a_{j_1 j_1 0} > 0, \alpha_{j_1 j_1} A_{j_2 j_2 1} > 0$, the flow at $y_{j_1}^* = a_{j_1 j_1 1} - b_{j_1 j_1 1}$ in the $y_{j_1}^{\perp}$-direction is a (SO, pF)-positive source flow.
- If $\alpha_{j_1 j_1} a_{j_1 j_1 0} < 0, \alpha_{j_1 j_1} A_{j_2 j_2 1} > 0$, the flow at $y_{j_1}^* = a_{j_1 j_1 1} - b_{j_1 j_1 1}$ in the $y_{j_1}^{\perp}$-direction is a (SI, pF)-positive sink flow.
- If $\alpha_{j_1 j_1} a_{j_1 j_1 0} > 0, \alpha_{j_1 j_1} A_{j_2 j_2 1} < 0$, the flow at $y_{j_1}^* = a_{j_1 j_1 1} - b_{j_1 j_1 1}$ in the $y_{j_1}^{\perp}$-direction is a (SO, nF)-negative source flow.
- If $\alpha_{j_1 j_1} a_{j_1 j_1 0} < 0, \alpha_{j_1 j_1} A_{j_2 j_2 1} < 0$, the flow at $y_{j_1}^* = a_{j_1 j_1 1} - b_{j_1 j_1 1}$ in the $y_{j_1}^{\perp}$-direction is a (SI, nF)-negative sink flow.

(ii$_2$) The flow of $y_{j_1}^* = a_{j_1 j_1 1}$ for the *two-dimensional dynamical system has the following properties:*

- *For* $\alpha_{j_1 j_2} a_{j_1 j_1 0} > 0, A_{j_2 j_2 1} > 0$,

$$(a_{j_1 j_1 1}, \dot{x}_{j_2}) = \underbrace{(\mathrm{UP}, \mathrm{pF})}_{\text{up-parabola flow } (+)} . \tag{1.16}$$

  The flow of $y_{j_1}^* = a_{j_1 j_1 1}$ *is an* (UP,pF)*-positive up-parabola flow.*
- *For* $\alpha_{j_1 j_2} a_{j_1 j_1 0} < 0, A_{j_2 j_2 1} > 0$,

$$(a_{j_1j_1 1}, \dot{x}_{j_2}) = \underbrace{(\mathrm{DP}, \mathrm{pF})}_{\text{down-parabola flow } (+)}. \tag{1.17}$$

The flow of $y^*_{j_1} = a_{j_1j_1 1}$ *is a* (DP,pF)-*positive down-parabola flow.*

- *For* $\alpha_{j_1j_2}a_{j_1j_1 0} > 0, A_{j_2j_2 1} < 0,$

$$(a_{j_1j_1 1}, \dot{x}_{j_2}) = \underbrace{(\mathrm{DP}, \mathrm{nF})}_{\text{down-parabola flow } (-)}. \tag{1.18}$$

The flow of $y^*_{j_1} = a_{j_1j_1 1}$ *is a* (DP,nF)-*negative down-parabola flow.*

- *For* $\alpha_{j_1j_2}a_{j_1j_1 0} < 0, A_{j_2j_2 1} < 0,$

$$(a_{j_1j_1 1}, \dot{x}_{j_2}) = \underbrace{(\mathrm{UP}, \mathrm{nF})}_{\text{up-parabola flow } (-)}. \tag{1.19}$$

The flow of $y^*_{j_1} = a_{j_1j_1 1}$ *is an* (UP,nF)-*negative up-parabola flow.*

(iii) If

$$A_{j_2j_2 1} \neq 0, A_{j_2j_2 0} \neq 0 \tag{1.20}$$

the standard form is

$$\begin{aligned} \dot{x}_{j_2} &= a_{j_2j_2 0}(y_{j_1} - a_{j_2j_2 1}), \\ \dot{x}_{j_1} &= a_{j_1j_1 0}(y_{j_1} - a_{j_1j_1 1}). \end{aligned} \tag{1.21}$$

There is a dynamical system in the y_{j_1}-direction *as*

$$\begin{aligned} &\dot{y}_{j_1} = b_{j_1j_1 0}(y_{j_1} - a_{j_1j_1 1} + b_{j_1j_1 1}); \\ &\text{with } b_{j_1j_1 0} = \alpha_{j_1j_1}a_{j_1j_1 0} + \alpha_{j_1j_2}a_{j_2j_2 0}, \\ &b_{j_1j_1 1} = \frac{\alpha_{j_1j_2}a_{j_2j_2 0}}{\alpha_{j_1j_1}a_{j_1j_1 0} + \alpha_{j_1j_2}a_{j_2j_2 0}}(a_{j_1j_1 1} - a_{j_2j_2 1}), \end{aligned} \tag{1.22}$$

and a dynamical system in the $y^{\perp}_{j_1}$*-direction as*

$$\begin{aligned} &\dot{y}^{\perp}_{j_1} = c_{j_1j_1 0}(y_{j_1} - a_{j_1j_1 1} + c_{j_1j_1 1}); \\ &\text{with } c_{j_1j_1 0} = \beta_{j_1j_1}a_{j_1j_1 0} + \beta_{j_1j_2}a_{j_2j_2 0}, \\ &c_{j_1j_1 1} = \frac{\beta_{j_1j_2}a_{j_2j_2 0}(a_{j_1j_1 1} - a_{j_2j_2 1})}{\beta_{j_1j_1}a_{j_1j_1 0} + \beta_{j_1j_2}a_{j_2j_2 0}} \end{aligned} \tag{1.23}$$

where

$$\begin{aligned} & y_{j_1}^{\perp} = \beta_{j_1 j_1} x_{j_1} + \beta_{j_1 j_2} x_{j_2}, \mathbf{n}_{j_1}^{\perp} = (\beta_{j_1 j_1}, \beta_{j_1 j_2})^{\mathrm{T}}; \\ & \mathbf{n}_{j_1} \cdot \mathbf{n}_{j_1}^{\perp} = \alpha_{j_1 j_1} \beta_{j_1 j_1} + \alpha_{j_1 j_2} \beta_{j_1 j_2} = 0. \end{aligned} \tag{1.24}$$

(iii₁) The flow along the y_{j_1} *-direction is*

$$t - t_0 = \frac{1}{b_{j_1 j_1 0}} \ln \frac{|y_{j_1} - a_{j_1 j_1 1} + b_{j_1 j_1 1}|}{|y_{j_1 0} - a_{j_1 j_1 1} + b_{j_1 j_1 1}|} \tag{1.25}$$

with the equilibriums of $y_{j_1}^* = b_{j_1 j_1 1}$ in the y_1-direction.

- If $b_{j_1 j_1 0} > 0$ and $\dot{y}_{j_1}^{\perp} > 0$, the flow along $y_{j_1}^* = a_{j_1 j_1 1} - b_{j_1 j_1 1}$ is a (SO,pF)-*positive source flow.*
- If $b_{j_1 j_1 0} > 0$ and $\dot{y}_{j_1}^{\perp} < 0$, the flow along $y_{j_1}^* = a_{j_1 j_1 1} - b_{j_1 j_1 1}$ is a (SO,nF)-*positive source flow.*
- *If* $b_{j_1 j_1 0} < 0$ *and* $\dot{y}_{j_1}^{\perp} > 0$*, the flow along* $y_{j_1}^* = a_{j_1 j_1 1} - b_{j_1 j_1 1}$ is a (SI,pF)- *positive source flow.*
- *If* $b_{j_1 j_1 0} < 0$ *and* $\dot{y}_{j_1}^{\perp} < 0$*, the flow along* $y_{j_1}^* = a_{j_1 j_1 1} - b_{j_1 j_1 1}$ is a (SI,nF)- *positive source flow.*

(iii₂) For $y_{j_1}^* = a_{j_1 j_1 1}$ *with* $\dot{x}_{j_2}^* = a_{j_2 j_2 0}(a_{j_1 j_1 1} - a_{j_2 j_2 1})$*, the two-dimensional dynamical system has the following properties:*

- *For* $\alpha_{j_1 j_2} a_{j_1 j_1 0} > 0, a_{j_2 j_2 0}(a_{j_1 j_1 1} - a_{j_2 j_2 1}) > 0,$

$$(a_{j_1 j_1 1}, x_{j_2}^*) = \underbrace{(\mathrm{UP}, \mathrm{pF})}_{\text{up-parabola flow } (+)}. \tag{1.26}$$

- *For* $\alpha_{j_1 j_2} a_{j_1 j_1 0} < 0, a_{j_2 j_2 0}(a_{j_1 j_1 1} - a_{j_2 j_2 1}) > 0,$

$$(a_{j_1 j_1 1}, \dot{x}_{j_2}^*) = \underbrace{(\mathrm{DP}, \mathrm{pF})}_{\text{down-parabola flow } (+)}. \tag{1.27}$$

- *For* $\alpha_{j_1 j_2} a_{j_1 j_1 0} > 0, a_{j_2 j_2 0}(a_{j_1 j_1 1} - a_{j_2 j_2 1}) < 0,$

$$(a_{j_1 j_1 1}, \dot{x}_{j_2}^*) = \underbrace{(\mathrm{DP}, \mathrm{nF})}_{\text{down-parabola flow } (-)}. \tag{1.28}$$

- *For* $\alpha_{j_1 j_2} a_{j_1 j_1 0} < 0, a_{j_2 j_2 0}(a_{j_1 j_1 1} - a_{j_2 j_2 1}) < 0,$

$$(a_{j_1 j_1 1}, \dot{x}_{j_2}^*) = \underbrace{(\mathrm{UP}, \mathrm{nF})}_{\text{up-parabola flow } (-)}. \tag{1.29}$$

(iii₃) For $y_{j_1}^* = a_{j_2j_2}1$ *with* $\dot{x}_{j_1}^* = a_{j_1j_1}0(a_{j_2j_2}1 - a_{j_1j_1}1)$, *the two-dimensional dynamical system has the following properties:*

- *For* $a_{j_1j_1}0(a_{j_2j_2}1 - a_{j_1j_1}1) > 0, \alpha_{j_1j_1}a_{j_2j_2}0 > 0$,

$$(\dot{x}_{j_1}^*, a_{j_2j_2}1) = \underbrace{(\text{pF}, \text{UP})}_{\text{up-parabola flow } (+)}. \tag{1.30}$$

- *For* $a_{j_1j_1}0(a_{j_2j_2}1 - a_{j_1j_1}1) < 0, \alpha_{j_1j_1}a_{j_2j_2}0 > 0$,

$$(\dot{x}_{j_1}^*, a_{j_2j_2}1) = \underbrace{(\text{nF}, \text{DP})}_{\text{down-parabola flow } (-)}. \tag{1.31}$$

- *For* $a_{j_1j_1}0(a_{j_2j_2}1 - a_{j_1j_1}1) > 0, \alpha_{j_1j_1}a_{j_2j_2}0 < 0$,

$$(\dot{x}_{j_1}^*, a_{j_2j_2}1) = \underbrace{(\text{pF}, \text{DP})}_{\text{down-parabola flow } (+)}. \tag{1.32}$$

- *For* $a_{j_1j_1}0(a_{j_2j_2}1 - a_{j_1j_1}1) < 0, \alpha_{j_1j_1}a_{j_2j_2}0 < 0$,

$$(\dot{x}_{j_1}^*, a_{j_2j_2}1) = \underbrace{(\text{nF}, \text{UP})}_{\text{up-parabola flow } (-)}. \tag{1.33}$$

(iii₄) For $a_{j_2j_2}1 = a_{j_1j_1}1$, *the standard form is*

$$\begin{aligned} \dot{x}_{j_2} &= a_{j_2j_2}0(y_{j_1} - a_{j_1j_1}1), \\ \dot{x}_{j_1} &= a_{j_1j_1}0(y_{j_1} - a_{j_1j_1}1) \end{aligned} \tag{1.34}$$

with the first integral manifold as

$$a_{j_1j_1}0(x_{j_2} - x_{j_20}) = a_{j_2j_2}0(x_{j_1} - x_{j_10}). \tag{1.35}$$

There is a dynamical system in the y_{j_1}*-direction as*

$$\dot{y}_{j_1} = b_{j_1j_1}0(y_{j_1} - a_{j_1j_1}1); \tag{1.36}$$

and a dynamical system in the $y_{j_1}^{\perp}$*-direction as*

$$\dot{y}_{j_1}^{\perp} = c_{j_1j_1}0(y_{j_1} - a_{j_1j_1}1). \tag{1.37}$$

The flow along the y_{j_1} *-direction is*

$$t - t_0 = \frac{1}{b_{j_1 j_1 0}} \ln \frac{|y_{j_1} - a_{j_1 j_1 1}|}{|y_{j_1 0} - a_{j_1 j_1 1}|} \tag{1.38}$$

with the equilibriums of $y^*_{j_1} = b_{j_1 j_1 1}$ *in the* y_1*-direction.*

Define a function as

$$G_{j_1 j_1 0} = a_{j_1 j_1 0} a_{j_1 j_1 0} + \alpha_{12} a_{j_2 j_2 0}. \tag{1.39}$$

The two-dimensional dynamical system possesses the infinite-equilibriums.

- *For* $a_{j_1 j_1 0}/a_{j_2 j_2 0} > 0$ *and* $G_{j_1 j_1 0} > 0$,

$$(a_{j_1 j_1 1}, a_{j_2 j_2 1}) = \underbrace{(\text{II}, \text{SO})}_{\text{increasing-inflection source}}. \tag{1.40}$$

- *For* $a_{j_1 j_1 0}/a_{j_2 j_2 0} > 0$ *and* $G_{j_1 j_1 0} < 0$,

$$(a_{j_1 j_1 1}, a_{j_2 j_2 1}) = \underbrace{(\text{II}, \text{SI})}_{\text{increasing-inflection sink}}. \tag{1.41}$$

- *For* $a_{j_1 j_1 0}/a_{j_2 j_2 0} < 0$ *and* $G_{j_1 j_1 0} > 0$,

$$(a_{j_1 j_1 1}, a_{j_2 j_2 1}) = \underbrace{(\text{DI}, \text{SO})}_{\text{decreasing-inflection source}}. \tag{1.42}$$

- *For* $a_{j_1 j_1 0}/a_{j_2 j_2 0} < 0$ *and* $G_{j_1 j_1 0} < 0$,

$$(a_{j_1 j_1 1}, a_{j_2 j_2 1}) = \underbrace{(\text{DI}, \text{SI})}_{\text{decreasing-inflection source}}. \tag{1.43}$$

(iii$_5$) For $a_{j_2 j_2 1} = a_{j_1 j_1 1}$, *the switching bifurcations for* $\alpha_{j_1 j_1} > 0$ *are given as follows.*

- *For* $a_{j_1 j_1 0} > 0$ and $a_{j_2 j_2 0} > 0$,

$$\left\{ \begin{array}{c} \underbrace{(\text{nF}, \text{DP})}_{\text{down-parabola flow } (-)} \\ \underbrace{(\text{UP}, \text{pF})}_{\text{up-parabola flow } (+)} \end{array} \right\} \rightleftharpoons \underbrace{(\text{II}, \text{SO})}_{\text{switching bifurcation}} \rightleftharpoons \left\{ \begin{array}{c} \underbrace{(\text{DP}, \text{nF})}_{\text{down-parabola flow } (-)} \\ \underbrace{(\text{pF}, \text{UP})}_{\text{up-parabola flow } (+)} \end{array} \right\}$$

for $\alpha_{j_1 j_2} > 0, G_{j_1 j_1 0} > 0$;

$$\begin{aligned}
&\left\{\begin{array}{c}\underbrace{(\mathrm{nF},\mathrm{DP})}_{\text{down-parabola flow }(-)}\\ \underbrace{(\mathrm{DP},\mathrm{pF})}_{\text{down-parabola flow }(+)}\end{array}\right\}\rightleftharpoons\underbrace{(\mathrm{II},\mathrm{SO})}_{\text{switching bifurcation}}\rightleftharpoons\left\{\begin{array}{c}\underbrace{(\mathrm{UP},\mathrm{nF})}_{\text{up-parabola flow }(-)}\\ \underbrace{(\mathrm{pF},\mathrm{UP})}_{\text{up-parabola flow }(+)}\end{array}\right\}\\
&\text{for } \alpha_{j_1j_2}<0, G_{j_1j_10}>0;\\
&\left\{\begin{array}{c}\underbrace{(\mathrm{nF},\mathrm{DP})}_{\text{down-parabola flow }(-)}\\ \underbrace{(\mathrm{DP},\mathrm{pF})}_{\text{down-parabola flow }(+)}\end{array}\right\}\rightleftharpoons\underbrace{(\mathrm{II},\mathrm{SI})}_{\text{switching bifurcation}}\rightleftharpoons\left\{\begin{array}{c}\underbrace{(\mathrm{UP},\mathrm{nF})}_{\text{up-parabola flow }(-)}\\ \underbrace{(\mathrm{pF},\mathrm{UP})}_{\text{up-parabola flow }(+)}\end{array}\right\}\\
&\text{for } \alpha_{j_1j_2}<0, G_{j_1j_10}<0.
\end{aligned}\tag{1.44}$$

- *For* $a_{j_1j_10}<0$ and $a_{j_2j_20}>0$,

$$\begin{aligned}
&\left\{\begin{array}{c}\underbrace{(\mathrm{pF},\mathrm{UP})}_{\text{up-parabola flow }(+)}\\ \underbrace{(\mathrm{DP},\mathrm{pF})}_{\text{down-parabola flow }(+)}\end{array}\right\}\rightleftharpoons\underbrace{(\mathrm{DI},\mathrm{SI})}_{\text{switching bifurcation}}\rightleftharpoons\left\{\begin{array}{c}\underbrace{(\mathrm{UP},\mathrm{nF})}_{\text{up-parabola flow }(-)}\\ \underbrace{(\mathrm{nF},\mathrm{DP})}_{\text{down-parabola flow }(-)}\end{array}\right\}\\
&\text{for } \alpha_{j_1j_2}>0, G_{j_1j_10}<0;\\
&\left\{\begin{array}{c}\underbrace{(\mathrm{pF},\mathrm{UP})}_{\text{up-parabola flow }(+)}\\ \underbrace{(\mathrm{DP},\mathrm{pF})}_{\text{down-parabola flow }(+)}\end{array}\right\}\rightleftharpoons\underbrace{(\mathrm{DI},\mathrm{SO})}_{\text{switching bifurcation}}\rightleftharpoons\left\{\begin{array}{c}\underbrace{(\mathrm{UP},\mathrm{nF})}_{\text{up-parabola flow }(-)}\\ \underbrace{(\mathrm{nF},\mathrm{DP})}_{\text{down-parabola flow }(-)}\end{array}\right\}\\
&\text{for } \alpha_{j_1j_2}>0, G_{j_1j_10}>0;\\
&\left\{\begin{array}{c}\underbrace{(\mathrm{pF},\mathrm{UP})}_{\text{up-parabola flow }(+)}\\ \underbrace{(\mathrm{UP},\mathrm{pF})}_{\text{up-parabola flow }(+)}\end{array}\right\}\rightleftharpoons\underbrace{(\mathrm{DI},\mathrm{SI})}_{\text{siwthcing bifurcation}}\rightleftharpoons\left\{\begin{array}{c}\underbrace{(\mathrm{DP},\mathrm{nF})}_{\text{down-parabola flow }(-)}\\ \underbrace{(\mathrm{nF},\mathrm{DP})}_{\text{down-parabola flow }(-)}\end{array}\right\}\\
&\text{for } \alpha_{j_1j_2}<0, G_{j_1j_10}<0.
\end{aligned}\tag{1.45}$$

- *For* $a_{j_1j_1 0} > 0$ and $a_{j_2j_2 0} < 0$,

$$\left\{\begin{array}{c}\underbrace{(\text{nF},\text{UP})}_{\text{up-parabola flow }(-)}\\ \underbrace{(\text{DP},\text{nF})}_{\text{down-parabola flow }(-)}\end{array}\right\} \rightleftharpoons \underbrace{(\text{DI},\text{SO})}_{\text{switching bifurcartion}} \rightleftharpoons \left\{\begin{array}{c}\underbrace{(\text{UP},\text{pF})}_{\text{up-parabola flow }(+)}\\ \underbrace{(\text{pF},\text{DP})}_{\text{down-parabola flow }(+)}\end{array}\right\}$$

for $\alpha_{j_1j_2} > 0, G_{j_1j_1 0} > 0$;

$$\left\{\begin{array}{c}\underbrace{(\text{nF},\text{UP})}_{\text{up-parabola flow }(-)}\\ \underbrace{(\text{DP},\text{nF})}_{\text{down-parabola flow }(-)}\end{array}\right\} \rightleftharpoons \underbrace{(\text{DI},\text{SI})}_{\text{switching bifurcartion}} \rightleftharpoons \left\{\begin{array}{c}\underbrace{(\text{UP},\text{pF})}_{\text{up-parabola flow }(+)}\\ \underbrace{(\text{pF},\text{DP})}_{\text{down-parabola flow }(+)}\end{array}\right\} \quad (1.46)$$

for $\alpha_{j_1j_2} > 0, G_{j_1j_1 0} < 0$;

$$\left\{\begin{array}{c}\underbrace{(\text{nF},\text{UP})}_{\text{up-parabola flow }(-)}\\ \underbrace{(\text{UP},\text{nF})}_{\text{up-parabola flow }(-)}\end{array}\right\} \rightleftharpoons \underbrace{(\text{DI},\text{SO})}_{\text{switching bifurcation}} \rightleftharpoons \left\{\begin{array}{c}\underbrace{(\text{DP},\text{pF})}_{\text{down-parabola flow }(+)}\\ \underbrace{(\text{pF},\text{DP})}_{\text{down-parabola flow }(+)}\end{array}\right\}$$

for $\alpha_{j_1j_2} < 0, G_{j_1j_1 0} > 0$.

- *For* $a_{j_1j_1 0} < 0$ and $a_{j_2j_2 0} < 0$,

$$\left\{\begin{array}{c}\underbrace{(\text{pF},\text{DP})}_{\text{down-parabola flow }(+)}\\ \underbrace{(\text{UP},\text{nF})}_{\text{up-parabola flow }(-)}\end{array}\right\} \rightleftharpoons \underbrace{(\text{II},\text{SO})}_{\text{switching bifurcation}} \rightleftharpoons \left\{\begin{array}{c}\underbrace{(\text{DP},\text{pF})}_{\text{down-parabola flow }(+)}\\ \underbrace{(\text{nF},\text{UP})}_{\text{up-parabola flow }(-)}\end{array}\right\}$$

for $\alpha_{j_1j_2} > 0, G_{j_1j_1 0} < 0$;

$$\left\{\begin{array}{c}\underbrace{(\text{pF},\text{DP})}_{\text{down-parabola flow }(+)}\\ \underbrace{(\text{DP},\text{nF})}_{\text{down-parabola flow }(-)}\end{array}\right\} \rightleftharpoons \underbrace{(\text{II},\text{SI})}_{\text{switching bifurcation}} \rightleftharpoons \left\{\begin{array}{c}\underbrace{(\text{UP},\text{pF})}_{\text{up-parabola flow }(+)}\\ \underbrace{(\text{nF},\text{UP})}_{\text{up-parabola flow }(-)}\end{array}\right\} \quad (1.47)$$

for $\alpha_{j_1j_2} < 0, G_{j_1j_1 0} < 0$;

$$\left\{\begin{array}{c}\underbrace{(\text{pF},\text{DP})}_{\text{down-parabola flow }(+)}\\ \underbrace{(\text{DP},\text{nF})}_{\text{down-parabola flow }(-)}\end{array}\right\} \rightleftharpoons \underbrace{(\text{II},\text{SO})}_{\text{switching bifurcation}} \rightleftharpoons \left\{\begin{array}{c}\underbrace{(\text{UP},\text{pF})}_{\text{up-parabola flow }(+)}\\ \underbrace{(\text{nF},\text{UP})}_{\text{up-parabola flow }(-)}\end{array}\right\}$$

for $\alpha_{j_1j_2} < 0, G_{j_1j_1 0} > 0$.

1.2 Proof of Theorem 1.1

(i) Consider

$$\begin{aligned}\dot{x}_{j_2} &= 0,\\ \dot{x}_{j_1} &= a_{j_1 j_1 0}(y_{j_1} - a_{j_1 j_1 1}).\end{aligned}$$

A new dynamical system in the y_{j_1}-direction is

$$\dot{y}_{j_1} = \alpha_{j_1 j_1} a_{j_1 j_1 0}(y_{j_1} - a_{j_1 j_1 1}),$$

and a dynamical system in the $y_{j_1}^{\perp}$-direction is

$$\dot{y}_{j_1}^{\perp} = \beta_{j_1 j_1} a_{j_1 j_1 0}(y_{j_1} - a_{j_1 j_1 1}),$$

where

$$\begin{aligned}&y_{j_1}^{\perp} = \beta_{j_1 j_1} x_{j_1} + \beta_{j_1 j_2} x_{j_2}, \mathbf{n}_{j_1}^{\perp} = (\beta_{j_1 j_1}, \beta_{j_1 j_2})^{\mathrm{T}};\\ &\mathbf{n}_{j_1} \cdot \mathbf{n}_{j_1}^{\perp} = \alpha_{j_1 j_1}\beta_{j_1 j_1} + \alpha_{j_1 j_2}\beta_{j_1 j_2} = 0.\end{aligned}$$

Through integration, the flow along the y_{j_1} -direction is obtained as

$$t - t_0 = \frac{1}{\alpha_{j_1 j_1} a_{j_1 j_1 0}} \ln \frac{|y_{j_1} - a_{j_1 j_1 1}|}{|y_{j_1 0} - a_{j_1 j_1 1}|}.$$

The variational equation at $y_{j_1}^{*} = a_{j_1 j_1 1}$ is

$$\Delta \dot{y}_{j_1} = \alpha_{j_1 j_1} a_{j_1 j_1 0} \Delta y_{j_1}.$$

The flows at $y_{j_1}^{*} = a_{j_1 j_1 1}$ are source and sink for $\alpha_{j_1 j_1} a_{j_1 j_1 0} > 0$ and $\alpha_{j_1 j_1} a_{j_1 j_1 0} < 0$, respectively.

Therefore, from the solutions, the following conclusions are obtained.

- For $a_{j_1 j_1 0}\alpha_{j_1 j_1} > 0$, the line of $y_{j_1}^{*} = a_{j_1 j_1 1}$ in the x_{j_1} -direction is a source infinite-equilibrium.
- If $a_{j_1 j_1 0}\alpha_{j_1 j_1} < 0$, the line of $y_{j_1}^{*} = a_{j_1 j_1 1}$ in the x_{j_1}-direction is a sink infinite-equilibrium.

(ii) If

$$A_{j_2 j_2 1} \neq 0, A_{j_2 j_2 0} = 0,$$

consider

$$\begin{aligned}\dot{x}_{j_2} &= A_{j_2 j_2 1},\\ \dot{x}_{j_1} &= a_{j_1 j_1 0}(y_{j_1} - a_{j_1 j_1 1}).\end{aligned}$$

(ii_1) The corresponding dynamical system in the y_{j_1}-direction is

$$\begin{aligned}\dot{y}_{j_1} &= \alpha_{j_1 j_2} A_{j_2 j_2 1} + \alpha_{j_1 j_1} a_{j_1 j_1 0}(y_{j_1} - a_{j_1 j_1 1})\\ &= \alpha_{j_1 j_1} a_{j_1 j_1 0}(y_{j_1} - a_{j_1 j_1 1} + b_{j_1 j_1 1}),\end{aligned}$$

and a dynamical system in the $y_{j_1}^{\perp}$-direction is

$$\dot{y}_{j_1}^{\perp} = \beta_{j_1 j_2} A_{j_2 j_2 1} + \beta_{j_1 j_1} a_{j_1 j_1 0}(y_{j_1} - a_{j_1 j_1 1}),$$

and

$$\begin{aligned}&y_{j_1}^{\perp} = \beta_{j_1 j_1} x_{j_1} + \beta_{j_1 j_2} x_{j_2}, \mathbf{n}_{j_1}^{\perp} = (\beta_{j_1 j_1}, \beta_{j_1 j_2})^{\mathrm{T}};\\ &\mathbf{n}_{j_1} \cdot \mathbf{n}_{j_1}^{\perp} = \alpha_{j_1 j_1} \beta_{j_1 j_1} + \alpha_{j_1 j_2} \beta_{j_1 j_2} = 0;\\ &\frac{\alpha_{j_1 j_2} A_{j_2 j_2 1}}{\alpha_{j_1 j_1} a_{j_1 j_1 0}} \equiv b_{j_1 j_1 1}.\end{aligned}$$

For $y_{j_1}^{*} = a_{j_1 j_1 1} - b_{j_1 j_1 1}$, we have

$$\dot{y}_{j_1}^{\perp *} = \frac{A_{j_2 j_2 1}}{\alpha_{j_1 j_1}}(\alpha_{j_1 j_2}^2 + \alpha_{j_1 j_2}^2).$$

Through integration, the flow along the y_{j_1} -direction is

$$t - t_0 = \frac{1}{\alpha_{j_1 j_1} a_{j_1 j_1 0}} \ln \frac{|y_{j_1} - a_{j_1 j_1 1} + b_{j_1 j_1 1}|}{|y_{j_1 0} - a_{j_1 j_1 1} + b_{j_1 j_1 1}|},$$

with the equilibriums of $y_{j_1}^{*} = a_{j_1 j_1 1} - b_{j_1 j_1 1}$ in the y_1-direction.

The variational equation at $y_{j_1}^{*} = a_{j_1 j_1 1} - b_{j_1 j_1 1}$ is

$$\Delta \dot{y}_{j_1} = \alpha_{j_1 j_1} a_{j_1 j_1 0} \Delta y_{j_1}.$$

The flows at $y_{j_1}^{*} = a_{j_1 j_1 1} - b_{j_1 j_1 1}$ are source and sink for $\alpha_{j_1 j_1} a_{j_1 j_1 0} > 0$ and $\alpha_{j_1 j_1} a_{j_1 j_1 0} < 0$, respectively.

Therefore, from the solution of y_{j_1} and $\dot{y}_{j_1}^{\perp *}$, we have the following conclusions.

- If $\alpha_{j_1j_1}a_{j_1j_1 0} > 0, \alpha_{j_1j_1}A_{j_2j_2 1} > 0$, the flow at $y^*_{j_1} = a_{j_1j_1 1} - b_{j_1j_1 1}$ in the $y^{\perp}_{j_1}$-direction is an (SO, pF)-positive source flow.
- If $\alpha_{j_1j_1}a_{j_1j_1 0} < 0, \alpha_{j_1j_1}A_{j_2j_2 1} > 0$, the flow at $y^*_{j_1} = a_{j_1j_1 1} - b_{j_1j_1 1}$ in the $y^{\perp}_{j_1}$-direction is an (SI, pF)-positive sink flow.
- If $\alpha_{j_1j_1}a_{j_1j_1 0} > 0, \alpha_{j_1j_1}A_{j_2j_2 1} < 0$, the flow at $y^*_{j_1} = a_{j_1j_1 1} - b_{j_1j_1 1}$ in the $y^{\perp}_{j_1}$-direction is an (SO, nF)-negative source flow.
- If $\alpha_{j_1j_1}a_{j_1j_1 0} < 0, \alpha_{j_1j_1}A_{j_2j_2 1} < 0$, the flow at $y^*_{j_1} = a_{j_1j_1 1} - b_{j_1j_1 1}$ in the $y^{\perp}_{j_1}$-direction is an (SI, nF)-negative sink flow.

(ii_2) Consider

$$\begin{aligned}\dot{x}_{j_2} &= A_{j_2j_2 1},\\ \dot{x}_{j_1} &= a_{j_1j_1 0}(y_{j_1} - a_{j_1j_1 1}).\end{aligned}$$

Thus, for $y^*_{j_1} = a_{j_1j_1 1}$, we have

$$\frac{dx_{j_1}}{dx_{j_2}}\bigg|_{y^*_{j_1} = a_{j_1j_1 1}} = \frac{a_{j_1j_1 0}(y_{j_1} - a_{j_1j_1 1})}{A_{j_2j_2 1}}\bigg|_{y^*_{j_1} = a_{j_1j_1 1}} = 0.$$

If

$$\frac{d^2x_{j_1}}{dx^2_{j_2}}\bigg|_{y^*_{j_1} = a_{j_1j_1 1}} = \frac{a_{j_1j_1 0}\alpha_{j_1j_2}}{A_{j_2j_2 1}} > 0,$$

there is an up-parabola flow in the x_{j_1}-direction. If

$$\frac{d^2x_{j_1}}{dx^2_{j_2}}\bigg|_{y^*_{j_1} = a_{j_1j_1 1}} = \frac{a_{j_1j_1 0}\alpha_{j_1j_2}}{A_{j_2j_2 1}} < 0,$$

there is a down-parabola flow in the x_{j_1}-direction.

From $\dot{x}_{j_2} = A_{j_2j_2 1}$, the flow at $y_{j_1} = a_{j_1j_1 1}$ is positive and negative on the x_{j_2} direction for $A_{j_2j_2 1} > 0$ and $A_{j_2j_2 1} < 0$, respectively.

Therefore, the flow of $y^*_{j_1} = a_{j_1j_1 1}$ for the two-dimensional dynamical system has the following properties:

- For $\alpha_{j_1j_2}a_{j_1j_1 0} > 0, A_{j_2j_2 1} > 0$,

$$(a_{j_1j_1 1}, \dot{x}_{j_2}) = \underbrace{(\text{UP}, \text{pF})}_{\text{up-parabola flow } (+)}.$$

- For $\alpha_{j_1j_2}a_{j_1j_1 0} < 0, A_{j_2j_2 1} > 0$,

$$(a_{j_1j_1}1, \dot{x}_{j_2}) = \underbrace{(\mathrm{DP}, \mathrm{pF})}_{\text{down-parabola flow } (+)}.$$

- For $\alpha_{j_1j_2}a_{j_1j_1}0 > 0, A_{j_2j_2}1 < 0$,

$$(a_{j_1j_1}1, \dot{x}_{j_2}) = \underbrace{(\mathrm{DP}, \mathrm{nF})}_{\text{down-parabola flow } (-)}.$$

- For $\alpha_{j_1j_2}a_{j_1j_1}0 < 0, A_{j_2j_2}1 < 0$,

$$(a_{j_1j_1}1, \dot{x}_{j_2}) = \underbrace{(\mathrm{UP}, \mathrm{nF})}_{\text{up-parabola flow } (-)}.$$

(iii) If

$$A_{j_2j_2}1 \neq 0, A_{j_2j_2}0 \neq 0,$$

consider

$$\begin{aligned}\dot{x}_{j_2} &= a_{j_2j_2}0(y_{j_1} - a_{j_2j_2}1),\\ \dot{x}_{j_1} &= a_{j_1j_1}0(y_{j_1} - a_{j_1j_1}1).\end{aligned}$$

There is a dynamical system in the y_{j_1}-direction as

$$\begin{aligned}&\dot{y}_{j_1} = b_{j_1j_1}0(y_{j_1} - a_{j_1j_1}1 + b_{j_1j_1}1);\\ &\text{with } b_{j_1j_1}0 = \alpha_{j_1j_1}a_{j_1j_1}0 + \alpha_{j_1j_2}a_{j_2j_2}0,\\ &b_{j_1j_1}1 = \frac{\alpha_{j_1j_2}a_{j_2j_2}0}{\alpha_{j_1j_1}a_{j_1j_1}0 + \alpha_{j_1j_2}a_{j_2j_2}0}(a_{j_1j_1}1 - a_{j_2j_2}1),\end{aligned}$$

and a dynamical system in the $y_{j_1}^{\perp}$-direction as

$$\begin{aligned}&\dot{y}_{j_1}^{\perp} = c_{j_1j_1}0(y_{j_1} - a_{j_1j_1}1 + c_{j_1j_1}1);\\ &\text{with } c_{j_1j_1}0 = \beta_{j_1j_1}a_{j_1j_1}0 + \beta_{j_1j_2}a_{j_2j_2}0,\\ &c_{j_1j_1}1 = \frac{\beta_{j_1j_2}a_{j_2j_2}0(a_{j_1j_1}1 - a_{j_2j_2}1)}{\beta_{j_1j_1}a_{j_1j_1}0 + \beta_{j_1j_2}a_{j_2j_2}0}\end{aligned}$$

where

$$y_{j_1}^{\perp} = \beta_{j_1 j_1} x_{j_1} + \beta_{j_1 j_2} x_{j_2}, \mathbf{n}_{j_1}^{\perp} = (\beta_{j_1 j_1}, \beta_{j_1 j_2})^{\mathrm{T}};$$
$$\mathbf{n}_{j_1} \cdot \mathbf{n}_{j_1}^{\perp} = \alpha_{j_1 j_1} \beta_{j_1 j_1} + \alpha_{j_1 j_2} \beta_{j_1 j_2} = 0.$$

The integration gives the flow based on the equilibriums of $y_{j_1}^* = a_{j_1 j_1 1} - b_{j_1 j_1 1}$ along the y_{j_1} -direction as

$$t - t_0 = \frac{1}{b_{j_1 j_1 0}} \ln \frac{|y_{j_1} - a_{j_1 j_1 1} + b_{j_1 j_1 1}|}{|y_{j_1 0} - a_{j_1 j_1 1} + b_{j_1 j_1 1}|}.$$

From the above solution of y_{j_1} with

$$\dot{y}_{j_1} = b_{j_1 j_1 0}(y_{j_1} - a_{j_1 j_1 1} + b_{j_1 j_1 1}),$$
$$\dot{y}_{j_1}^{\perp} = c_{j_1 j_1 0}(y_{j_1} - a_{j_1 j_1 1} + c_{j_1 j_1 1}),$$

the following conclusions are obtained.

- If $b_{j_1 j_1 0} > 0$ and $\dot{y}_{j_1}^{\perp} > 0$, the flow along $y_{j_1}^* = a_{j_1 j_1 1} - b_{j_1 j_1 1}$ is a (SO,pF)-positive source flow.
- If $b_{j_1 j_1 0} > 0$ and $\dot{y}_{j_1}^{\perp} < 0$, the flow along $y_{j_1}^* = a_{j_1 j_1 1} - b_{j_1 j_1 1}$ is a (SO,nF)-positive source flow.
- If $b_{j_1 j_1 0} < 0$ and $\dot{y}_{j_1}^{\perp} > 0$, the flow along $y_{j_1}^* = a_{j_1 j_1 1} - b_{j_1 j_1 1}$ is a (SI,pF)- positive source flow.
- If $b_{j_1 j_1 0} < 0$ and $\dot{y}_{j_1}^{\perp} < 0$, the flow along $y_{j_1}^* = a_{j_1 j_1 1} - b_{j_1 j_1 1}$ is a (SI,nF)- positive source flow.

(iii$_2$) Consider

$$\dot{x}_{j_2} = a_{j_2 j_2 0}(y_{j_1} - a_{j_2 j_2 1}),$$
$$\dot{x}_{j_1} = a_{j_1 j_1 0}(y_{j_1} - a_{j_1 j_1 1}).$$

Thus, for $y_{j_1}^* = a_{j_1 j_1 1}$,

$$\frac{dx_{j_1}}{dx_{j_2}}\Big|_{y_{j_1}^* = a_{j_1 j_1 1}} = \frac{a_{j_1 j_1 0}(y_{j_1} - a_{j_1 j_1 1})}{a_{j_2 j_2 0}(y_{j_1} - a_{j_2 j_2 1})}\Big|_{y_{j_1}^* = a_{j_1 j_1 1}} = 0.$$

If

$$\frac{d^2 x_{j_1}}{dx_{j_2}^2}\Big|_{y_{j_1}^* = a_{j_1 j_1 1}} = \frac{a_{j_1 j_1 0} \alpha_{j_1 j_2}}{a_{j_2 j_2 0}(a_{j_1 j_1 1} - a_{j_2 j_2 1})} > 0,$$

there is an up-parabola flow in the x_{j_1}-direction. If

$$\frac{d^2 x_{j_1}}{dx_{j_2}^2}\Big|_{y_{j_1}^*=a_{j_1j_1 1}} = \frac{a_{j_1j_1 0}\alpha_{j_1j_2}}{a_{j_2j_2 0}(a_{j_1j_1 1}-a_{j_2j_2 1})} < 0,$$

there is a down-parabola flow in the x_{j_1}-direction. From $\dot{x}_{j_2}^* = a_{j_2j_2 0}(a_{j_1j_1 1}-a_{j_2j_2 1})$, the flow at $y_{j_1} = a_{j_1j_1 1}$ is positive and negative on the x_{j_2} direction for $\dot{x}_{j_2}^* > 0$ and $\dot{x}_{j_2}^* < 0$, respectively.

Therefore, for $y_1^* = a_{j_1j_1 1}$ with $\dot{x}_{j_2}^* = a_{j_2j_2 0}(a_{j_1j_1 1}-a_{j_2j_2 1})$, the two-dimensional dynamical system has the following properties:

- For $\alpha_{j_1j_2}a_{j_1j_1 0} > 0, a_{j_2j_2 0}(a_{j_1j_1 1}-a_{j_2j_2 1}) > 0$,

$$(a_{j_1j_1 1}, \dot{x}_{j_2}^*) = \underbrace{(\text{UP}, \text{pF})}_{\text{up-parabola flow } (+)}.$$

- For $\alpha_{j_1j_2}a_{j_1j_1 0} < 0, a_{j_2j_2 0}(a_{j_1j_1 1}-a_{j_2j_2 1}) > 0$,

$$(a_{j_1j_1 1}, \dot{x}_{j_2}^*) = \underbrace{(\text{DP}, \text{pF})}_{\text{down-parabola flow } (+)}.$$

- For $\alpha_{j_1j_2}a_{j_1j_1 0} > 0, a_{j_2j_2 0}(a_{j_1j_1 1}-a_{j_2j_2 1}) < 0$,

$$(a_{j_1j_1 1}, \dot{x}_{j_2}^*) = \underbrace{(\text{DP}, \text{nF})}_{\text{down-parabola flow } (-)}.$$

- For $\alpha_{j_1j_2}a_{j_1j_1 0} < 0, a_{j_2j_2 0}(a_{j_1j_1 1}-a_{j_2j_2 1}) < 0$,

$$(a_{j_1j_1 1}, \dot{x}_{j_2}^*) = \underbrace{(\text{UP}, \text{nF})}_{\text{up-parabola flow } (-)}.$$

(iii_3) Similarly, consider

$$\begin{aligned}\dot{x}_{j_2} &= a_{j_2j_2 0}(y_{j_1} - a_{j_2j_2 1}),\\ \dot{x}_{j_1} &= a_{j_1j_1 0}(y_{j_1} - a_{j_1j_1 1}).\end{aligned}$$

Thus, for $y_{j_1}^* = a_{j_2j_2 1}$,

$$\frac{dx_{j_2}}{dx_{j_1}}\Big|_{y_{j_1}^*=a_{j_2j_2 1}} = \frac{a_{j_2j_2 0}(y_{j_1} - a_{j_2j_2 1})}{a_{j_1j_1 0}(y_{j_1} - a_{j_1j_1 1})}\Big|_{y_{j_1}^*=a_{j_2j_2 1}} = 0.$$

If

$$\left.\frac{d^2 x_{j_2}}{dx_{j_1}^2}\right|_{y_{j_1}^*=a_{j_2j_2 1}} = \frac{a_{j_2j_2 0}\alpha_{j_1j_1}}{a_{j_1j_1 0}(a_{j_2j_2 1} - a_{j_1j_1 1})} > 0,$$

there is an up-parabola curve for x_{j_2}-direction. If

$$\left.\frac{d^2 x_{j_2}}{dx_{j_1}^2}\right|_{y_{j_1}^*=a_{j_2j_2 1}} = \frac{a_{j_2j_2 0}\alpha_{j_1j_1}}{a_{j_1j_1 0}(a_{j_2j_2 1} - a_{j_1j_1 1})} < 0,$$

there is a down-parabola curve for x_{j_2}-direction. From $\dot{x}_{j_1}^* = a_{j_2j_2 0}(a_{j_2j_2 1} - a_{j_1j_1 1})$, the flow at $y_{j_1}^* = a_{j_2j_2 1}$ is positive and negative on the x_{j_1} direction for $\dot{x}_{j_1}^* > 0$ and $\dot{x}_{j_1}^* < 0$, respectively.

Therefore, for $y_1^* = a_{j_1j_1 2}$ with $\dot{x}_{j_1}^* = a_{j_1j_1 0}(a_{j_2j_2 1} - a_{j_1j_1 1})$, the two-dimensional dynamical system has the following properties:

- For $a_{j_1j_1 0}(a_{j_2j_2 1} - a_{j_1j_1 1}) > 0, \alpha_{j_1j_1}a_{j_2j_2 0} > 0$,

$$(\dot{x}_{j_1}^*, a_{j_2j_2 1}) = \underbrace{(\text{pF}, \text{UP})}_{\text{up-parabola flow }(+)}.$$

- For $a_{j_1j_1 0}(a_{j_2j_2 1} - a_{j_1j_1 1}) < 0, \alpha_{j_1j_1}a_{j_2j_2 0} > 0$,

$$(\dot{x}_{j_1}^*, a_{j_2j_2 1}) = \underbrace{(\text{nF}, \text{DP})}_{\text{down-parabola flow }(-)}.$$

- For $a_{j_1j_1 0}(a_{j_2j_2 1} - a_{j_1j_1 1}) > 0, \alpha_{j_1j_1}a_{j_2j_2 0} < 0$,

$$(\dot{x}_{j_1}^*, a_{j_2j_2 1}) = \underbrace{(\text{pF}, \text{DP})}_{\text{down-parabola flow }(+)}.$$

- For $a_{j_1j_1 0}(a_{j_2j_2 1} - a_{j_1j_1 1}) < 0, \alpha_{j_1j_1}a_{j_2j_2 0} < 0$,

$$(\dot{x}_{j_1}^*, a_{j_2j_2 1}) = \underbrace{(\text{nF}, \text{UP})}_{\text{up-parabola flow }(-)}.$$

(iii$_4$) For $a_{j_2j_2 1} = a_{j_1j_1 1}$, consider the dynamical system as

$$\begin{aligned}\dot{x}_{j_2} &= a_{j_2j_2 0}(y_{j_1} - a_{j_1j_1 1}),\\ \dot{x}_{j_1} &= a_{j_1j_1 0}(y_{j_1} - a_{j_1j_1 1}).\end{aligned}$$

Thus, for $y_{j_1}^* \neq a_{j_1j_1 1}$ we have

$$\frac{dx_{j_2}}{dx_{j_1}} = \frac{a_{j_1j_10}}{a_{j_2j_20}}.$$

and the first integral manifold is

$$a_{j_1j_10}(x_{j_2} - x_{j_20}) = a_{j_2j_20}(x_{j_1} - x_{j_10}).$$

which is a straight line in phase space to be intersected with $y_{j_1} = a_{j_1j_11}$.

Similarly, construct a dynamical system in the y_{j_1}-direction as

$$\dot{y}_{j_1} = b_{j_1j_10}(y_{j_1} - a_{j_1j_11});$$

and a dynamical system in the $y_{j_1}^{\perp}$-direction is also constructed by

$$\dot{y}_{j_1}^{\perp} = c_{j_1j_10}(y_{j_1} - a_{j_1j_11}).$$

The flow along the y_{j_1} -direction is

$$t - t_0 = \frac{1}{b_{j_1j_10}} \ln \frac{|y_{j_1} - a_{j_1j_11}|}{|y_{j_10} - a_{j_1j_11}|},$$

with the equilibriums of $y_{j_1}^* = b_{j_1j_11}$ in the y_1-direction.

For $y_{j_1}^* = a_{j_1j_11} = a_{j_2j_21}$, if

$$\frac{dx_{j_1}}{dx_{j_2}}\Big|_{y_{j_1}^* = a_{j_1j_1}} = \frac{a_{j_1j_10}}{a_{j_2j_20}} > 0,$$

there is an increasing-inflection flow for $y_{j_1}^* = a_{j_1j_11} = a_{j_2j_21}$; and if

$$\frac{dx_{j_1}}{dx_{j_2}}\Big|_{y_{j_1}^* = a_{j_1j_1}} = \frac{a_{j_1j_10}}{a_{j_2j_20}} < 0,$$

there is a decreasing-inflection flow for $y_{j_1}^* = a_{j_1j_11} = a_{j_2j_21}$.

In the vicinity of $y_{j_1}^* = a_{j_1j_11} = a_{j_2j_21}$, we have

$$\Delta\dot{y}_{j_1} = (\alpha_{j_1j_1} a_{j_1j_10} + \alpha_{j_1j_2} a_{j_2j_20})\Delta y_{j_1} = G_{j_1j_10}\Delta y_{j_1},$$

with

$$G_{j_1j_10} = \alpha_{j_1j_1} a_{j_1j_10} + \alpha_{j_1j_2} a_{j_2j_20}.$$

The inflection flows are source and sink for $G_{j_1j_10} > 0$ and $G_{j_1j_10} < 0$.

Thus, the two-dimensional dynamical system possesses the infinite-equilibriums with the following properties as in Eqs. (1.40)–(1.43).

- For $a_{j_1j_10}/a_{j_2j_20} > 0$ and $G_{j_1j_10} > 0$,

$$(a_{j_1j_11}, a_{j_2j_21}) = \underbrace{(\mathrm{II}, \mathrm{SO})}_{\text{increasing-inflection source}}.$$

- For $a_{j_1j_10}/a_{j_2j_20} > 0$ and $G_{j_1j_10} < 0$,

$$(a_{j_1j_11}, a_{j_2j_21}) = \underbrace{(\mathrm{II}, \mathrm{SI})}_{\text{increasing-inflection sink}}.$$

- For $a_{j_1j_10}/a_{j_2j_20} < 0$ and $G_{j_1j_10} > 0$,

$$(a_{j_1j_11}, a_{j_2j_21}) = \underbrace{(\mathrm{DI}, \mathrm{SO})}_{\text{decreasing-inflection source}}.$$

- For $a_{j_1j_10}/a_{j_2j_20} < 0$ and $G_{j_1j_10} < 0$,

$$(a_{j_1j_11}, a_{j_2j_21}) = \underbrace{(\mathrm{DI}, \mathrm{SI})}_{\text{decreasing-inflection source}}.$$

(iii$_5$) For $a_{j_2j_21} = a_{j_1j_11}$, the switching bifurcations are determined from the cases of (iii$_2$)–(iii$_4$). Thus, the corresponding switching bifurcations for $\alpha_{j_1j_1} > 0$ are presented as in Eqs. (1.44)–(1.47).

- For $a_{j_1j_10} > 0$ and $a_{j_2j_20} > 0$,

$$\left\{\begin{array}{c}\underbrace{(\mathrm{nF}, \mathrm{DP})}_{\text{down-parabola flow }(-)}\\ \underbrace{(\mathrm{UP}, \mathrm{pF})}_{\text{up-parabola flow }(+)}\end{array}\right\} \rightleftharpoons \underbrace{(\mathrm{II}, \mathrm{SO})}_{\text{switching bifurcation}} \rightleftharpoons \left\{\begin{array}{c}\underbrace{(\mathrm{DP}, \mathrm{nF})}_{\text{down-parabola flow }(-)}\\ \underbrace{(\mathrm{pF}, \mathrm{UP})}_{\text{up-parabola flow }(+)}\end{array}\right\}$$

for $\alpha_{j_1j_2} > 0, G_{j_1j_10} > 0$;

$$\left\{\begin{array}{c}\underbrace{(\mathrm{nF},\mathrm{DP})}_{\text{down-parabola flow }(-)}\\ \underbrace{(\mathrm{DP},\mathrm{pF})}_{\text{down-parabola flow }(+)}\end{array}\right\}\rightleftharpoons\underbrace{(\mathrm{II},\mathrm{SO})}_{\text{switching bifurcation}}\rightleftharpoons\left\{\begin{array}{c}\underbrace{(\mathrm{UP},\mathrm{nF})}_{\text{up-parabola flow }(-)}\\ \underbrace{(\mathrm{pF},\mathrm{UP})}_{\text{up-parabola flow }(+)}\end{array}\right\}$$
for $\alpha_{j_1j_2} < 0, G_{j_1j_10} > 0$;

$$\left\{\begin{array}{c}\underbrace{(\mathrm{nF},\mathrm{DP})}_{\text{down-parabola flow }(-)}\\ \underbrace{(\mathrm{DP},\mathrm{pF})}_{\text{down-parabola flow }(+)}\end{array}\right\}\rightleftharpoons\underbrace{(\mathrm{II},\mathrm{SI})}_{\text{switching bifurcation}}\rightleftharpoons\left\{\begin{array}{c}\underbrace{(\mathrm{UP},\mathrm{nF})}_{\text{up-parabola flow }(-)}\\ \underbrace{(\mathrm{pF},\mathrm{UP})}_{\text{up-parabola flow }(+)}\end{array}\right\}$$
for $\alpha_{j_1j_2} < 0, G_{j_1j_10} < 0$.

- For $a_{j_1j_10} < 0$ and $a_{j_2j_20} > 0$,

$$\left\{\begin{array}{c}\underbrace{(\mathrm{pF},\mathrm{UP})}_{\text{up-parabola flow }(+)}\\ \underbrace{(\mathrm{DP},\mathrm{pF})}_{\text{down-parabola flow }(+)}\end{array}\right\}\rightleftharpoons\underbrace{(\mathrm{DI},\mathrm{SI})}_{\text{switching bifurcation}}\rightleftharpoons\left\{\begin{array}{c}\underbrace{(\mathrm{UP},\mathrm{nF})}_{\text{up-parabola flow }(-)}\\ \underbrace{(\mathrm{nF},\mathrm{DP})}_{\text{down-parabola flow }(-)}\end{array}\right\}$$
for $\alpha_{j_1j_2} > 0, G_{j_1j_10} < 0$;

$$\left\{\begin{array}{c}\underbrace{(\mathrm{pF},\mathrm{UP})}_{\text{up-parabola flow }(+)}\\ \underbrace{(\mathrm{DP},\mathrm{pF})}_{\text{down-parabola flow }(+)}\end{array}\right\}\rightleftharpoons\underbrace{(\mathrm{DI},\mathrm{SO})}_{\text{switching bifurcation}}\rightleftharpoons\left\{\begin{array}{c}\underbrace{(\mathrm{UP},\mathrm{nF})}_{\text{up-parabola flow }(-)}\\ \underbrace{(\mathrm{nF},\mathrm{DP})}_{\text{down-parabola flow }(-)}\end{array}\right\}$$
for $\alpha_{j_1j_2} > 0, G_{j_1j_10} > 0$;

$$\left\{\begin{array}{c}\underbrace{(\mathrm{pF},\mathrm{UP})}_{\text{up-parabola flow }(+)}\\ \underbrace{(\mathrm{UP},\mathrm{pF})}_{\text{up-parabola flow }(+)}\end{array}\right\}\rightleftharpoons\underbrace{(\mathrm{DI},\mathrm{SI})}_{\text{siwthcing bifurcation}}\rightleftharpoons\left\{\begin{array}{c}\underbrace{(\mathrm{DP},\mathrm{nF})}_{\text{down-parabola flow }(-)}\\ \underbrace{(\mathrm{nF},\mathrm{DP})}_{\text{down-parabola flow }(-)}\end{array}\right\}$$
for $\alpha_{j_1j_2} < 0, G_{j_1j_10} < 0$.

- For $a_{j_1j_10} > 0$ and $a_{j_2j_20} < 0$,

$$\left\{\begin{array}{c}\underbrace{(\mathrm{nF},\mathrm{UP})}_{\text{up-parabola flow }(-)}\\ \underbrace{(\mathrm{DP},\mathrm{nF})}_{\text{down-parabola flow }(-)}\end{array}\right\}\rightleftharpoons\underbrace{(\mathrm{DI},\mathrm{SO})}_{\text{switching bifurcartion}}\rightleftharpoons\left\{\begin{array}{c}\underbrace{(\mathrm{UP},\mathrm{pF})}_{\text{up-parabola flow }(+)}\\ \underbrace{(\mathrm{pF},\mathrm{DP})}_{\text{down-parabola flow }(+)}\end{array}\right\}$$
for $\alpha_{j_1j_2} > 0, G_{j_1j_10} > 0$;

$$\left\{\begin{array}{c}\underbrace{(\mathrm{nF},\mathrm{UP})}_{\text{up-parabola flow }(-)}\\ \underbrace{(\mathrm{DP},\mathrm{nF})}_{\text{down-parabola flow }(-)}\end{array}\right\} \rightleftharpoons \underbrace{(\mathrm{DI},\mathrm{SI})}_{\text{switching bifurcartion}} \rightleftharpoons \left\{\begin{array}{c}\underbrace{(\mathrm{UP},\mathrm{pF})}_{\text{up-parabola flow }(+)}\\ \underbrace{(\mathrm{pF},\mathrm{DP})}_{\text{down-parabola flow }(+)}\end{array}\right\}$$
for $\alpha_{j_1j_2} > 0, G_{j_1j_10} < 0$;

$$\left\{\begin{array}{c}\underbrace{(\mathrm{nF},\mathrm{UP})}_{\text{up-parabola flow }(-)}\\ \underbrace{(\mathrm{UP},\mathrm{nF})}_{\text{up-parabola flow }(-)}\end{array}\right\} \rightleftharpoons \underbrace{(\mathrm{DI},\mathrm{SO})}_{\text{switching bifurcation}} \rightleftharpoons \left\{\begin{array}{c}\underbrace{(\mathrm{DP},\mathrm{pF})}_{\text{down-parabola flow }(+)}\\ \underbrace{(\mathrm{pF},\mathrm{DP})}_{\text{down-parabola flow }(+)}\end{array}\right\}$$
for $\alpha_{j_1j_2} < 0, G_{j_1j_10} > 0$.

- For $a_{j_1j_10} < 0$ and $a_{j_2j_20} < 0$,

$$\left\{\begin{array}{c}\underbrace{(\mathrm{pF},\mathrm{DP})}_{\text{down-parabola flow }(+)}\\ \underbrace{(\mathrm{UP},\mathrm{nF})}_{\text{up-parabola flow }(-)}\end{array}\right\} \rightleftharpoons \underbrace{(\mathrm{II},\mathrm{SO})}_{\text{switching bifurcation}} \rightleftharpoons \left\{\begin{array}{c}\underbrace{(\mathrm{DP},\mathrm{pF})}_{\text{down-parabola flow }(+)}\\ \underbrace{(\mathrm{nF},\mathrm{UP})}_{\text{up-parabola flow }(-)}\end{array}\right\}$$
for $\alpha_{j_1j_2} > 0, G_{j_1j_10} < 0$;

$$\left\{\begin{array}{c}\underbrace{(\mathrm{pF},\mathrm{DP})}_{\text{down-parabola flow }(+)}\\ \underbrace{(\mathrm{DP},\mathrm{nF})}_{\text{down-parabola flow }(-)}\end{array}\right\} \rightleftharpoons \underbrace{(\mathrm{II},\mathrm{SI})}_{\text{switching bifurcation}} \rightleftharpoons \left\{\begin{array}{c}\underbrace{(\mathrm{UP},\mathrm{pF})}_{\text{up-parabola flow }(+)}\\ \underbrace{(\mathrm{nF},\mathrm{UP})}_{\text{up-parabola flow }(-)}\end{array}\right\}$$
for $\alpha_{j_1j_2} < 0, G_{j_1j_10} < 0$;

$$\left\{\begin{array}{c}\underbrace{(\mathrm{pF},\mathrm{DP})}_{\text{down-parabola flow }(+)}\\ \underbrace{(\mathrm{DP},\mathrm{nF})}_{\text{down-parabola flow }(-)}\end{array}\right\} \rightleftharpoons \underbrace{(\mathrm{II},\mathrm{SO})}_{\text{switching bifurcation}} \rightleftharpoons \left\{\begin{array}{c}\underbrace{(\mathrm{UP},\mathrm{pF})}_{\text{up-parabola flow }(+)}\\ \underbrace{(\mathrm{nF},\mathrm{UP})}_{\text{up-parabola flow }(-)}\end{array}\right\}$$
for $\alpha_{j_1j_2} < 0, G_{j_1j_10} > 0$.

Therefore, this theorem is proved. ■

1.3 One-Dimensional Flows and Switching Bifurcations

In this section, the one-dimensional flows and switching bifurcations will be presented.

1.3.1 Source and Sink Infinite-Equilibriums

Consider a two-dimensional system as

$$\begin{aligned}\dot{x}_1 &= a_{110}(y_1 - a_{111}),\\ \dot{x}_2 &= A_{221} = 0,\end{aligned} \tag{1.48}$$

where

$$y_1 = \alpha_{11}x_1 + \alpha_{12}x_2. \tag{1.49}$$

There is a dynamical system in the y_1-direction as

$$\dot{y}_1 = \alpha_1 a_{110}(y_1 - a_{111}), \tag{1.50}$$

with a dynamical system in the $y_1^{\perp}$-direction as

$$\dot{y}_1^{\perp} = \beta_{11}a_{110}(y_1 - a_{111}), \tag{1.51}$$

where

$$\begin{aligned}&y_1^{\perp} = \beta_{11}x_1 + \beta_{12}x_2, \mathbf{n}_1^{\perp} = (\beta_{11}, \beta_{12})^{\mathrm{T}};\\ &\mathbf{n}_1 \cdot \mathbf{n}_1^{\perp} = \alpha_1\beta_{11} + \alpha_{12}\beta_{12} = 0,\end{aligned} \tag{1.52}$$

then the flow along the y_1 -direction is

$$t - t_0 = \frac{1}{\alpha_{11}a_{110}} \ln \frac{|y_1 - a_{111}|}{|y_{10} - a_{111}|}. \tag{1.53}$$

If $\alpha_{12} = 0$, y_1 is equivalent to x_1. Such a system is an independent system in the x_1-direction. If $\alpha_{11} = 0$, y_1 is equivalent to x_2. Such a system becomes a constant speed system, which is a crossing-variable system. The two-dimensional system has an infinite sink and source equilibrium of $y_1 = a_{111}$ in the x_1-direction, as shown in Fig. 1.1a, b for $a_{110}\alpha_{11} > 0$ and $a_{110}\alpha_{11} < 0$, respectively.

1.3.2 Parabola Flows

Consider a two-dimensional system as

$$\begin{aligned}\dot{x}_1 &= a_{110}(y_1 - a_{111}),\\ \dot{x}_2 &= A_{221} \neq 0.\end{aligned} \tag{1.54}$$

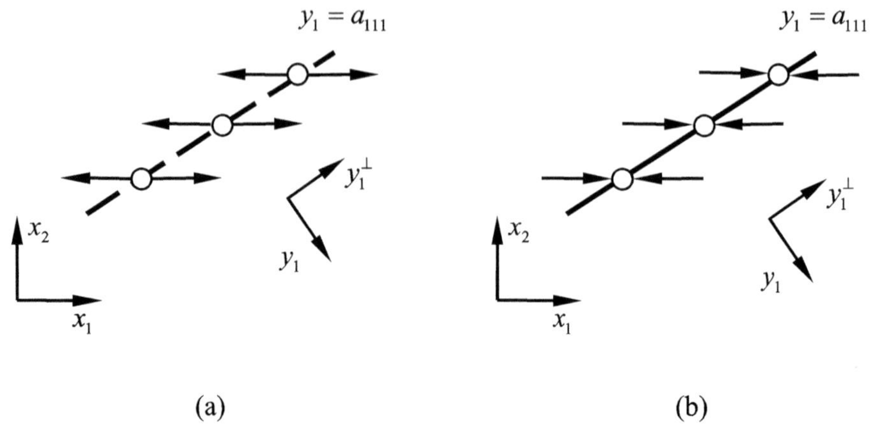

Fig. 1.1 Phase portraits for two-dimensional systems in the x_1-direction with $y_1^* = a_{111}$ and static on the x_2-direction with $\dot{x}_2 = A_{221} = 0$: (**a**) an infinite source equilibrium on the line of $y_1^* = a_{111}$ in the x_1-direction for $(a_{110}\alpha_{11} > 0, A_{221} = 0)$, (**b**) an infinite sink equilibrium on the line of $y_1^* = a_{111}$ in the x_1-direction $(a_{110}\alpha_{11} < 0, A_{221} = 0)$. $(\alpha_{11} > 0, \alpha_{12} < 0)$

Thus

$$\begin{aligned} \alpha_{11}(x_1 - x_{10}) &= -\frac{\alpha_{12}A_{221}}{\alpha_{11}a_{110}} \ln \frac{|\,\alpha_{11}a_{110}(y_1 - a_{111}) + \alpha_{12}A_{221}\,|}{|\,\alpha_{11}a_{110}(y_{10} - a_{111}) + \alpha_{12}A_{221}\,|} + (y_1 - y_{10}), \\ x_2 - x_{20} &= \frac{A_{221}}{\alpha_{11}a_{110}} \ln \frac{|\,\alpha_{11}a_{110}(y_1 - a_{111}) + \alpha_{12}A_{221}\,|}{|\,\alpha_{11}a_{110}(y_{10} - a_{111}) + \alpha_{12}A_{221}\,|}. \end{aligned} \tag{1.55}$$

On the other hand, we have

$$\dot{y}_1 = \alpha_{11}a_{110}(y_1 - a_{111}) + \alpha_{12}A_{221}, \tag{1.56}$$

and the solution is

$$t - t_0 = \frac{1}{a_{110}\alpha_{11}} \ln \frac{|\,y_1 - a_{111} + b_{111}\,|}{|\,y_{10} - a_{111} + b_{111}\,|} \text{ with } b_{111} = \frac{\alpha_{12}A_{221}}{\alpha_{11}a_{110}}, \tag{1.57}$$

and the equilibrium in the y_1 -direction is

$$y_1^* = a_{111} - b_{111}. \tag{1.58}$$

If $\alpha_{11}a_{110} > 0$, the equilibrium of $y_1^* = a_{111} - b_{111}$ in the y_1-direction is a source flow along the $y_1^{\perp}$-direction, and If $\alpha_{11}a_{110} < 0$, the equilibrium of $y_1^* = a_{111} - b_{111}$ in the y_1-direction is a sink flow along the $y_1^{\perp}$-direction.

The dynamics for the perpendicular coordinate $y_1^{\perp}$ is determined by

$$\dot{y}_1^{\perp} = \beta_{11} a_{110}(y_1 - a_{111}) + \beta_{12} A_{221}, \tag{1.59}$$

where

$$\begin{aligned} & y_1^{\perp} = \beta_{11} x_1 + \beta_{12} x_2, \mathbf{n}_1^{\perp} = (\beta_{11}, \beta_{12})^{\mathrm{T}}, \\ & (\mathbf{n}_1)^{\mathrm{T}} \cdot \mathbf{n}_1^{\perp} = \alpha_{11}\beta_{11} + \alpha_{12}\beta_{12} = 0 \Rightarrow \beta_{11} = -\alpha_{12}, \beta_{12} = \alpha_{11}. \end{aligned} \tag{1.60}$$

The corresponding equilibrium in the $y_1^{\perp}$-direction is

$$y_1^* = a_{111} - c_{111}, \text{with } c_{111} = \frac{\beta_{12} A_{221}}{\beta_{11} a_{110}}. \tag{1.61}$$

From the foregoing equations, the flow generated by Eq. (1.54) is the one-dimensional flow on the line of $y_1 = a_{111}$, as shown in Fig. 1.2. $y_1 = \alpha_{11} x_1 + \alpha_{12} x_2$ with $\alpha_{11} > 0$ and $\alpha_{12} < 0$. In Fig. 1.2a, the (DP,pF)-positive *down-parabola* flow on the line of $y_1^* = a_{111}$ ($a_{110} > 0$, $A_{221} > 0$) is presented for $a_{110}\alpha_{12} < 0$. Due to $\alpha_{11} a_{110} > 0$, the equilibrium of $y_1^* = a_{111} - b_{111}$ is a positive source flow. Due to $b_{111} < 0$, $y_1^* = a_{111} - b_{111} > a_{111}$. So the flow is moving along the positive $y_1^{\perp}$ direction with $\dot{y}_1^{\perp} > 0$. The (UP,pF)-positive *up-parabola* flow on the line of $y_1^* = a_{111}$ ($a_{110} < 0$, $A_{221} > 0$) is presented in Fig. 1.2b for $a_{110}\alpha_{12} > 0$. Due to $\alpha_{11} a_{110} > 0$, the equilibrium of $y_1^* = a_{111} - b_{111}$ is a positive sink flow. Due to $b_{111} > 0$, $y_1^* = a_{111} - b_{111} < a_{111}$. So the flow is moving along the positive $y_1^{\perp}$ direction with $\dot{y}_1^{\perp} > 0$. In Fig. 1.2c, the (UP,nF)-negative *up-parabola* flow on the line of $y_1^* = a_{111}$ ($a_{110} > 0$, $A_{221} < 0$) is presented for $a_{110}\alpha_{12} < 0$. Due to $\alpha_{11} a_{110} < 0$, the equilibrium of $y_1^* = a_{111} - b_{111}$ is a negative source flow. For $b_{111} > 0$, the flow of $y_1^* = a_{111} - b_{111} < a_{111}$ is moving along the positive $y_1^{\perp}$-direction with $\dot{y}_1^{\perp} < 0$. The (DP,nF)-negative *down-parabola* flow on the line of $y_1^* = a_{111}$ ($a_{110} < 0$, $A_{221} < 0$) is presented in Fig. 1.2d for $a_{110}\alpha_{12} > 0$. Due to $\alpha_{11} a_{110} < 0$, the equilibrium of $y_1^* = a_{111} - b_{111}$ is a negative sink flow. Due to $b_{111} < 0$, $y_1^* = a_{111} - b_{111} > a_{111}$. So the flow is moving along the negative $y_1^{\perp}$-direction with $\dot{y}_1^{\perp} < 0$.

For $\alpha_{11} > 0$ and $\alpha_{12} < 0$, we have

$$\left.\begin{aligned} & (a_{111}, \dot{x}_2) = \underbrace{(\mathrm{DP}, \mathrm{pF})}_{\text{flow}} \\ & (a_{111} - b_{111}, \dot{y}_1^{\perp}) = \underbrace{(\mathrm{SO}, \mathrm{pF})}_{\text{flow}} \end{aligned}\right\} \text{for } a_{110} > 0, A_{221} > 0; \tag{1.62}$$

$$\left.\begin{aligned} & (a_{111}, \dot{x}_2) = \underbrace{(\mathrm{UP}, \mathrm{pF})}_{\text{flow}} \\ & (a_{111} - b_{111}, \dot{y}_1^{\perp}) = \underbrace{(\mathrm{SI}, \mathrm{pF})}_{\text{flow}} \end{aligned}\right\} \text{for } a_{110} < 0, A_{221} > 0; \tag{1.63}$$

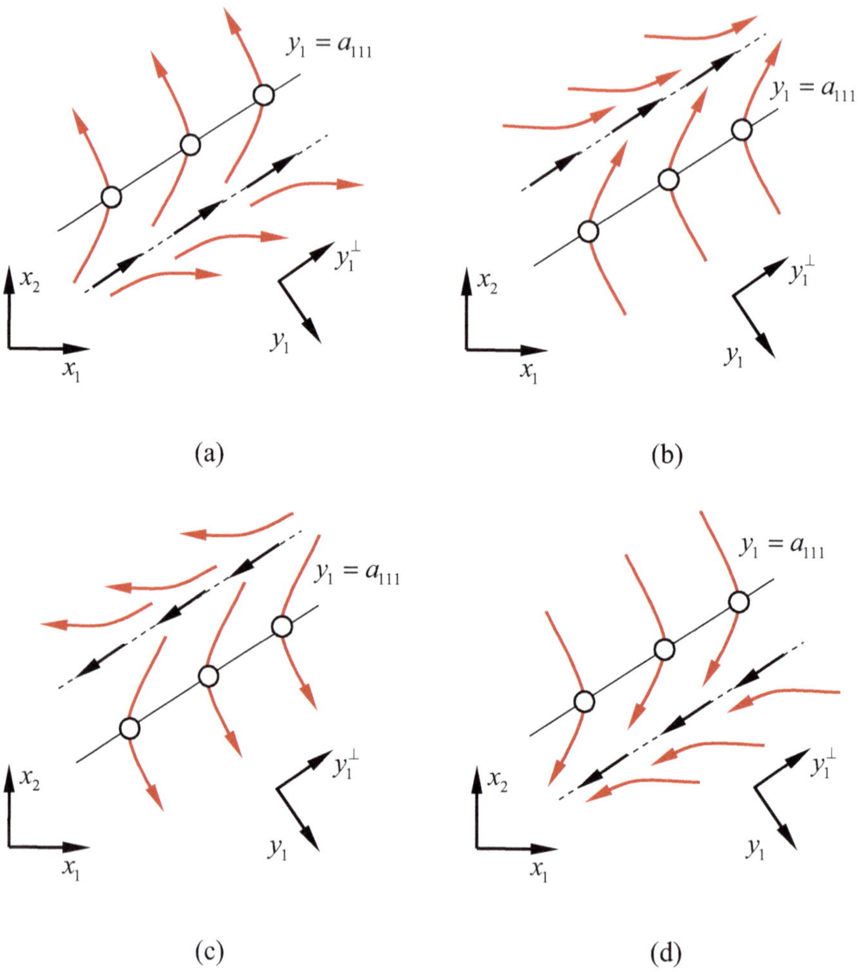

Fig. 1.2 Phase portraits for two-dimensional systems on the x_1-direction with $y_1^* = a_{111}$ and on the x_2-direction with $\dot{x}_2 = a_{221} \neq 0$: (**a**) (DP,pF)-positive down-parabola flow on the line of $y_1^* = a_{111}$ ($a_{110} > 0$, $A_{221} > 0$), (**b**) (UP,pF)-positive up-parabola flow on the line of $y_1^* = a_{111}$ ($a_{110} < 0$, $A_{221} > 0$), (**c**) (UP,nF)-negative up-parabola flow on the line of $y_1^* = a_{111}$ ($a_{110} > 0$, $A_{221} < 0$), (**d**) (DP,nF)-negative down-parabola flow on the line of $y_1^* = a_{111}$ ($a_{110} < 0$, $A_{221} < 0$). ($\alpha_{11} > 0$, $\alpha_{12} < 0$)

$$\left.\begin{aligned} (a_{111}, \dot{x}_2) &= \underbrace{(\mathrm{UP}, \mathrm{nF})}_{\text{flow}} \\ (a_{111} - b_{111}, \dot{y}_1^{\perp}) &= \underbrace{(\mathrm{SO}, \mathrm{nF})}_{\text{flow}} \end{aligned}\right\} \text{for } a_{110} > 0, A_{221} < 0, \tag{1.64}$$

$$\left.\begin{array}{l}(a_{111},\dot{x}_2)=\underbrace{(\mathrm{DP},\mathrm{nF})}_{\text{flow}},\\ (a_{111}-b_{111},\dot{y}_1^{\perp})=\underbrace{(\mathrm{SI},\mathrm{nF})}_{\text{flow}}\end{array}\right\}\text{ for }a_{110}<0,A_{221}<0. \tag{1.65}$$

From the foregoing equations, the flow generated by Eq. (1.54) is the one-dimensional flow on the line of $y_1=a_{111}$, as shown in Fig. 1.3. $y_1=\alpha_{11}x_1+\alpha_{12}x_2$ with $\alpha_{11}>0$ and $\alpha_{12}>0$. In Fig. 1.3a, the (UP,pF)-positive up-parabola flow on the line of $y_1^*=a_{111}$ ($a_{110}>0$, $A_{221}>0$) is presented for $a_{110}\alpha_{12}>0$. Owing to $\alpha_{11}a_{110}>0$, the equilibrium of $y_1^*=a_{111}-b_{111}$ is a positive source flow. Due to $b_{111}>0$, $y_1^*=a_{111}-b_{111}<a_{111}$. So the flow is moving along the positive $y_1^{\perp}$ direction with $\dot{y}_1^{\perp}>0$. The (DP,pF)-positive down-parabola flow on the line of $y_1^*=a_{111}$ ($a_{110}<0$, $A_{221}>0$) is presented in Fig. 1.3b for $a_{110}\alpha_{12}<0$. Owing to $\alpha_{11}a_{110}>0$, the equilibrium of $y_1^*=a_{111}-b_{111}$ is a positive sink flow. Due to $b_{111}<0$, $y_1^*=a_{111}-b_{111}>a_{111}$. So the flow is moving along the positive $y_1^{\perp}$ direction with $\dot{y}_1^{\perp}>0$. In Fig. 1.3c, the (DP,nF)-negative down-parabola flow on the line of $y_1^*=a_{111}$ ($a_{110}>0$, $A_{221}<0$) is presented for $a_{110}\alpha_{12}>0$. Owing to $\alpha_{11}a_{110}<0$, the equilibrium of $y_1^*=a_{111}-b_{111}$ is a negative source flow. For $y_1^*=a_{111}-b_{111}>a_{111}$ due to $b_{111}<0$, the flow is moving along the negative $y_1^{\perp}$-direction with $\dot{y}_1^{\perp}<0$. The (UP,nF)-negative up-parabola flow on the line of $y_1^*=a_{111}$ ($a_{110}<0$, $A_{221}<0$) is presented in Fig. 1.3d for $a_{110}\alpha_{12}<0$. Due to $\alpha_{11}a_{110}<0$, the equilibrium of $y_1^*=a_{111}-b_{111}$ is a negative sink flow. Due to $b_{111}>0$, $y_1^*=a_{111}-b_{111}<a_{111}$. So the flow is moving along the negative $y_1^{\perp}$-direction with $\dot{y}_1^{\perp}<0$.

For $\alpha_{11}>0$ and $\alpha_{12}>0$, we have

$$\left.\begin{array}{l}(a_{111},\dot{x}_2)=\underbrace{(\mathrm{UP},\mathrm{pF})}_{\text{flow}},\\ (a_{111}-b_{111},\dot{y}_1^{\perp})=\underbrace{(\mathrm{SO},\mathrm{pF})}_{\text{flow}}\end{array}\right\}\text{ for }a_{110}>0\text{ and }A_{221}>0; \tag{1.66}$$

$$\left.\begin{array}{l}(a_{111},\dot{x}_2)=\underbrace{(\mathrm{DP},\mathrm{pF})}_{\text{flow}},\\ (a_{111}-b_{111},\dot{y}_1^{\perp})=\underbrace{(\mathrm{SI},\mathrm{pF})}_{\text{flow}}\end{array}\right\}\text{ for }a_{110}<0\text{ and }A_{221}>0; \tag{1.67}$$

$$\left.\begin{array}{l}(a_{111},\dot{x}_2)=\underbrace{(\mathrm{DP},\mathrm{nF})}_{\text{flow}},\\ (a_{111}-b_{111},\dot{y}_1^{\perp})=\underbrace{(\mathrm{SO},\mathrm{nF})}_{\text{flow}}\end{array}\right\}\text{ for }a_{110}>0\text{ and }A_{221}<0; \tag{1.68}$$

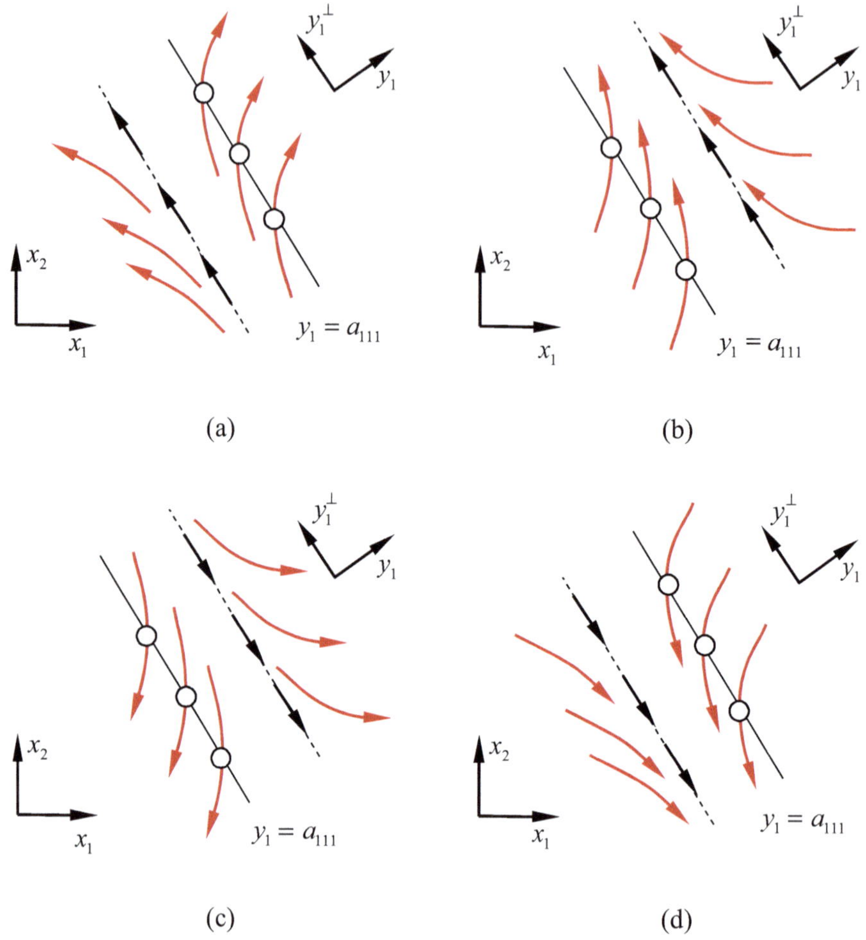

Fig. 1.3 Phase portraits for two-dimensional systems on the x_1-direction with $y_1^* = a_{111}$ and on the x_2-direction with $\dot{x}_2 = a_{221} \neq 0$: (**a**) (UP,pF)-positive up-parabola flow on the line of $y_1^* = a_{111}$ ($a_{110} > 0$, $A_{221} > 0$), (**b**) (DP,pF)-positive down-parabola flow on the line of $y_1^* = a_{111}$ ($a_{110} < 0$, $A_{221} > 0$), (**c**) (DP,nF)-negative down-parabola flow on the line of $y_1^* = a_{111}$ ($a_{110} > 0$, $A_{221} < 0$), (**d**) (UP,nF)-negative up-parabola flow on the line of $y_1^* = a_{111}$ ($a_{110} < 0$, $A_{221} < 0$). ($\alpha_{11} > 0$, $\alpha_{12} > 0$)

$$\left.\begin{aligned} (a_{111}, \dot{x}_2) &= \underbrace{(\mathrm{UP}, \mathrm{nF})}_{\text{flow}}, \\ (a_{111} - b_{111}, \dot{y}_1^{\perp}) &= \underbrace{(\mathrm{SI}, \mathrm{nF})}_{\text{flow}} \end{aligned}\right\} \text{for } a_{110} < 0 \text{ and } A_{221} < 0. \tag{1.69}$$

If $\alpha_{11} = 0$, such a flow is a perfect parabolic flow via

$$x_1 - x_{10} = \frac{1}{2}\frac{a_{110}}{\alpha_{12}A_{221}}\left[(\alpha_{12}x_2 - a_{111})^2 - (\alpha_{12}x_{20} - a_{111})^2\right]. \tag{1.70}$$

If $\alpha_{12} = 0$, such a flow is a perfect logarithmic flow through

$$x_2 - x_{20} = \frac{A_{221}}{\alpha_{11}a_{110}} \ln \frac{|\alpha_{11}x_1 - a_{111}|}{|\alpha_{11}x_{10} - a_{111}|}. \tag{1.71}$$

Consider a single bivariate linear system as

$$\begin{aligned} \dot{x}_2 &= a_{220}(y_1 - a_{221}), \\ \dot{x}_1 &= a_{110}(y_1 - a_{111}), \end{aligned} \tag{1.72}$$

where

$$y_1 = \alpha_{11}x_1 + \alpha_{12}x_2, \mathbf{n}_1 = (\alpha_{11}, \alpha_{12})^{\mathrm{T}}. \tag{1.73}$$

There is a dynamical system in the y_1-direction as

$$\begin{aligned} &\dot{y}_1 = b_{110}(y_1 - a_{111} + b_{111}); \\ &\text{with } b_{110} = \alpha_{11}a_{110} + \alpha_{12}a_{220}, \\ &b_{111} = \frac{\alpha_{12}a_{220}}{\alpha_{11}a_{110} + \alpha_{12}a_{220}}(a_{111} - a_{221}), \end{aligned} \tag{1.74}$$

and a dynamical system in the $y_1^{\perp}$-direction as

$$\begin{aligned} &\dot{y}_1^{\perp} = c_{110}(y_1 - a_{111} + c_{111}); \\ &\text{with } c_{110} = \beta_{11}a_{110} + \beta_{12}a_{220}, \\ &c_{111} = \frac{\beta_{12}a_{220}(a_{111} - a_{221})}{\beta_{11}a_{110} + \beta_{12}a_{220}}, \end{aligned} \tag{1.75}$$

where

$$\begin{aligned} &y_1^{\perp} = \beta_{11}x_1 + \beta_{12}x_2, \mathbf{n}_1^{\perp} = (\beta_{11}, \beta_{12})^{\mathrm{T}}; \\ &\mathbf{n}_1 \cdot \mathbf{n}_1^{\perp} = \alpha_{11}\beta_{11} + \alpha_{12}\beta_{12} = 0. \end{aligned} \tag{1.76}$$

Thus, for $a_{221} < a_{111}$ and ($\alpha_{11} > 0$, $\alpha_{12} > 0$), the two-dimensional dynamical system has the following properties:

$$\left\{ \begin{array}{c} (a_{111}, \dot{x}_2) \\ (\dot{x}_1, a_{221}) \end{array} \right\} = \left\{ \begin{array}{c} \underbrace{(\mathrm{UP}, \mathrm{pF})}_{\text{flow}} \\ \underbrace{(\mathrm{nF}, \mathrm{DP})}_{\text{flow}} \end{array} \right\} \text{ for } a_{110} > 0, a_{220} > 0; \tag{1.77}$$

$$\left\{\begin{array}{c}(a_{111},\dot{x}_2)\\(\dot{x}_1,a_{221})\end{array}\right\}=\left\{\begin{array}{c}\underbrace{(\mathrm{DP},\mathrm{pF})}_{\text{flow}}\\\underbrace{(\mathrm{pF},\mathrm{UP})}_{\text{flow}}\end{array}\right\}\text{for } a_{110}<0, a_{220}>0; \tag{1.78}$$

$$\left\{\begin{array}{c}(a_{111},\dot{x}_2)\\(\dot{x}_1,a_{221})\end{array}\right\}=\left\{\begin{array}{c}\underbrace{(\mathrm{DP},\mathrm{nF})}_{\text{flow}}\\\underbrace{(\mathrm{nF},\mathrm{UP})}_{\text{flow}}\end{array}\right\}\text{for } a_{110}>0, a_{220}<0; \tag{1.79}$$

$$\left\{\begin{array}{c}(a_{111},\dot{x}_2)\\(\dot{x}_1,a_{221})\end{array}\right\}=\left\{\begin{array}{c}\underbrace{(\mathrm{UP},\mathrm{nF})}_{\text{flow}}\\\underbrace{(\mathrm{pF},\mathrm{DP})}_{\text{flow}}\end{array}\right\}\text{for } a_{110}<0, a_{220}<0. \tag{1.80}$$

The corresponding phase portraits ($a_{111} > a_{221}$ and $\alpha_{11} > 0$, $\alpha_{12} < 0$) for two-dimensional systems on the x_1-direction with $y_1^* = a_{111}$ and on the x_2-direction with $y_1^* = a_{221}$ are presented in Fig. 1.4a–d. The one-dimensional source and sink flows for $y_1 = a_{111} - b_{111}$ are also presented.

For $a_{221} < a_{111}$ and ($\alpha_{11} > 0$, $\alpha_{12} < 0$), the two-dimensional dynamical system has the following properties:

$$\left\{\begin{array}{c}(a_{111},\dot{x}_2)\\(\dot{x}_1,a_{221})\end{array}\right\}=\left\{\begin{array}{c}\underbrace{(\mathrm{DP},\mathrm{pF})}_{\text{flow}}\\\underbrace{(\mathrm{nF},\mathrm{DP})}_{\text{flow}}\end{array}\right\}\text{for } a_{110}>0, a_{220}>0; \tag{1.81}$$

$$\left\{\begin{array}{c}(a_{111},\dot{x}_2)\\(\dot{x}_1,a_{221})\end{array}\right\}=\left\{\begin{array}{c}\underbrace{(\mathrm{UP},\mathrm{pF})}_{\text{flow}}\\\underbrace{(\mathrm{pF},\mathrm{UP})}_{\text{flow}}\end{array}\right\}\text{for } a_{110}<0, a_{220}>0; \tag{1.82}$$

$$\left\{\begin{array}{c}(a_{111},\dot{x}_2)\\(\dot{x}_1,a_{221})\end{array}\right\}=\left\{\begin{array}{c}\underbrace{(\mathrm{UP},\mathrm{nF})}_{\text{flow}}\\\underbrace{(\mathrm{nF},\mathrm{UP})}_{\text{flow}}\end{array}\right\}\text{for } a_{110}>0, a_{220}<0; \tag{1.83}$$

$$\left\{\begin{array}{c}(a_{111},\dot{x}_2)\\(\dot{x}_1,a_{221})\end{array}\right\}=\left\{\begin{array}{c}\underbrace{(\mathrm{DP},\mathrm{nF})}_{\text{flow}}\\\underbrace{(\mathrm{pF},\mathrm{DP})}_{\text{flow}}\end{array}\right\}\text{for } a_{110}<0, a_{220}<0. \tag{1.84}$$

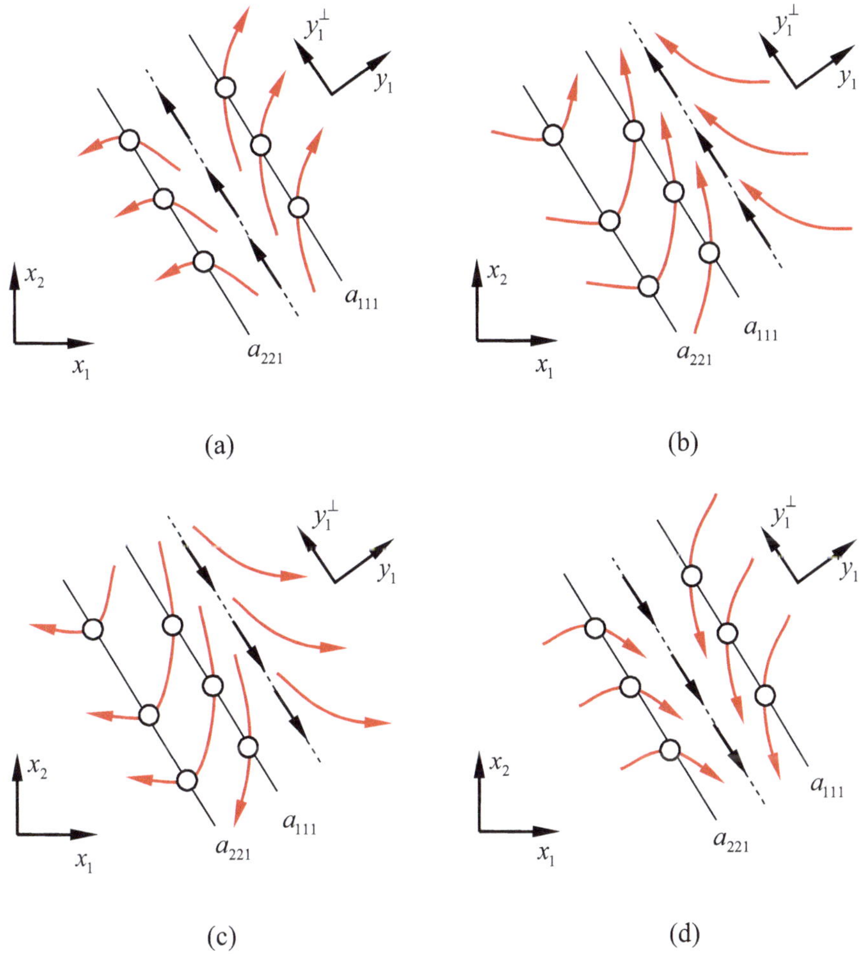

Fig. 1.4 Phase portraits ($a_{111} > a_{221}$ and $\alpha_{11} > 0, \alpha_{12} > 0$) for two-dimensional systems on the x_1-direction with $y_1^* = a_{111}$ and on the x_2-direction with $y_1^* = a_{221}$: (**a**) (UP,pF)-flow of $y_1^* = a_{111}$ and (nF,DP)-flow of $y_1^* = a_{221}$ ($a_{110} > 0$, $a_{220} > 0$), (**b**) (DP,pF)-flow of $y_1^* = a_{111}$ and (pF,UP)-flow of $y_1^* = a_{221}$ ($a_{110} < 0$, $a_{220} > 0$), (**c**) (DP,nF)-flow of $y_1^* = a_{111}$ and (nF,UP)-flow of $y_1^* = a_{221}$ ($a_{110} > 0$, $a_{220} < 0$), (**d**) (UP,nF)-flow of $y_1^* = a_{111}$ and (pF,DP)-flow of $y_1^* = a_{221}$ ($a_{110} < 0$, $a_{220} < 0$)

The corresponding phase portraits ($a_{111} > a_{221}$ and $\alpha_{11} > 0$, $\alpha_{12} < 0$) for two-dimensional systems on the x_1-direction with $y_1^* = a_{111}$ and on the x_2-direction with $y_1^* = a_{221}$ are presented in Fig. 1.5a–d. The one-dimensional source and sink flows for $y_1 = a_{111} - b_{111}$ are also presented.

For $a_{221} > a_{111}$ and ($\alpha_{11} > 0$, $\alpha_{12} > 0$), the two-dimensional dynamical system has the following one-dimensional flow.

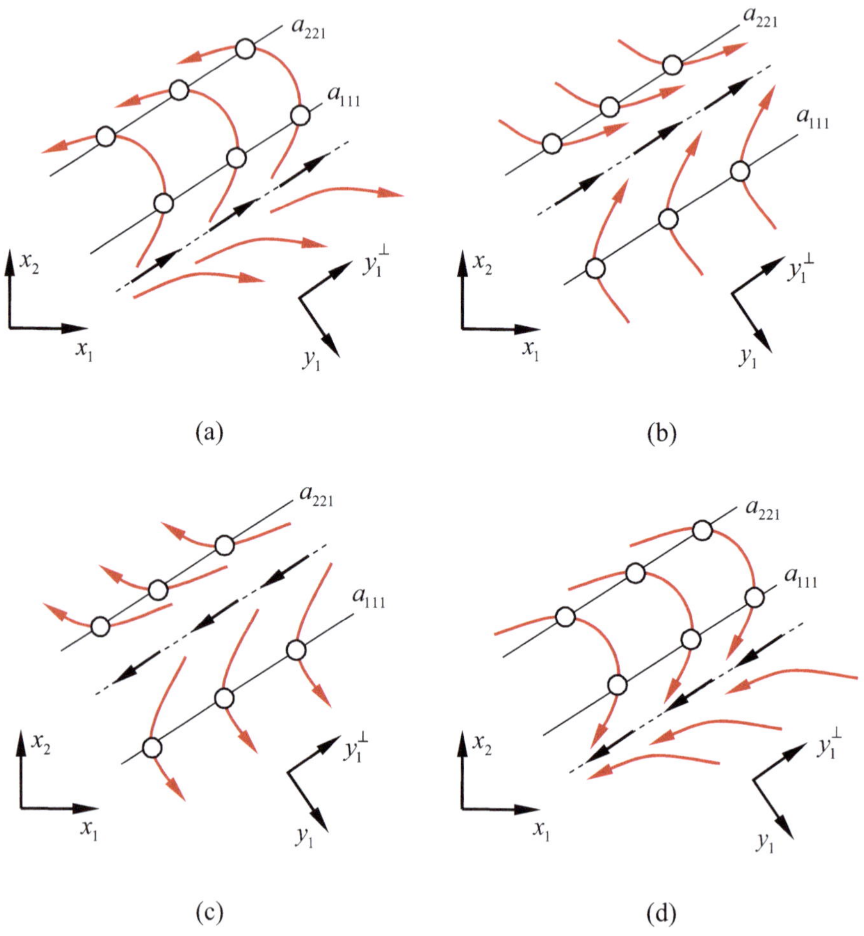

Fig. 1.5 Phase portraits ($a_{111} > a_{221}$ and $\alpha_{11} > 0$, $\alpha_{12} < 0$) for two-dimensional systems on the x_1-direction with $y_1^* = a_{111}$ and on the x_2-direction with $y_1^* = a_{221}$: (**a**) (DP,pF)-flow of $y_1^* = a_{111}$ and (nF,DP)-flow of $y_1^* = a_{221}$ ($a_{110} > 0$, $a_{220} > 0$), (**b**) (UP,pF)-flow of $y_1^* = a_{111}$ and (pF,UP)-flow of $y_1^* = a_{221}$ ($a_{110} < 0$, $a_{220} > 0$), (**c**) (UP,nF)-flow of $y_1^* = a_{111}$ and (nF,UP)-flow of $y_1^* = a_{221}$ ($a_{110} > 0$, $a_{220} < 0$), (**d**) (DP,nF)-flow of $y_1^* = a_{111}$ and (pF,DP)-flow of $y_1^* = a_{221}$ ($a_{110} < 0$, $a_{220} < 0$)

$$\left\{ \begin{array}{c} (a_{111}, \dot{x}_2) \\ (\dot{x}_1, a_{221}) \end{array} \right\} = \left\{ \begin{array}{c} \underbrace{(\mathrm{DP}, \mathrm{nF})}_{\text{flow}} \\ \underbrace{(\mathrm{pF}, \mathrm{UP})}_{\text{flow}} \end{array} \right\} \text{for } a_{110} > 0, a_{220} > 0; \tag{1.85}$$

$$\left\{ \begin{array}{c} (a_{111}, \dot{x}_2) \\ (\dot{x}_1, a_{221}) \end{array} \right\} = \left\{ \begin{array}{c} \underbrace{(\mathrm{UP}, \mathrm{nF})}_{\text{flow}} \\ \underbrace{(\mathrm{nF}, \mathrm{DP})}_{\text{flow}} \end{array} \right\} \text{ for } a_{110} < 0, a_{220} > 0; \tag{1.86}$$

$$\left\{\begin{array}{l}(a_{111},\dot{x}_2)\\(\dot{x}_1,a_{221})\end{array}\right\}=\left\{\begin{array}{c}\underbrace{(\mathrm{UP},\mathrm{pF})}_{\text{flow}}\\\underbrace{(\mathrm{pF},\mathrm{DP})}_{\text{flow}}\end{array}\right\}\text{ for } a_{110}<0, a_{220}>0; \tag{1.87}$$

$$\left\{\begin{array}{l}(a_{111},\dot{x}_2)\\(\dot{x}_1,a_{221})\end{array}\right\}=\left\{\begin{array}{c}\underbrace{(\mathrm{DP},\mathrm{pF})}_{\text{flow}}\\\underbrace{(\mathrm{nF},\mathrm{UP})}_{\text{flow}}\end{array}\right\}\text{ for } a_{110}<0, a_{220}<0. \tag{1.88}$$

The corresponding phase portraits ($a_{111} < a_{221}$ and $\alpha_{11} > 0$, $\alpha_{12} > 0$) for two-dimensional systems on the x_1-direction with $y_1^* = a_{111}$ and on the x_2-direction with $y_1^* = a_{221}$ are presented in Fig. 1.6a–d. The one-dimensional source and sink flows for $y_1 = a_{111} - b_{111}$ are also presented.

For $a_{221} > a_{111}$ and ($\alpha_{11} > 0$, $\alpha_{12} < 0$), the two-dimensional dynamical system has the following one-dimensional flow as follows.

$$\left\{\begin{array}{l}(a_{111},\dot{x}_2)\\(\dot{x}_1,a_{221})\end{array}\right\}=\left\{\begin{array}{c}\underbrace{(\mathrm{UP},\mathrm{nF})}_{\text{flow}}\\\underbrace{(\mathrm{pF},\mathrm{UP})}_{\text{flow}}\end{array}\right\}\text{ for } a_{110}>0, a_{220}>0; \tag{1.89}$$

$$\left\{\begin{array}{l}(a_{111},\dot{x}_2)\\(\dot{x}_1,a_{221})\end{array}\right\}=\left\{\begin{array}{c}\underbrace{(\mathrm{DP},\mathrm{nF})}_{\text{flow}}\\\underbrace{(\mathrm{nF},\mathrm{DP})}_{\text{flow}}\end{array}\right\}\text{ for } a_{110}<0, a_{220}>0; \tag{1.90}$$

$$\left\{\begin{array}{l}(a_{111},\dot{x}_2)\\(\dot{x}_1,a_{221})\end{array}\right\}=\left\{\begin{array}{c}\underbrace{(\mathrm{DP},\mathrm{pF})}_{\text{flow}}\\\underbrace{(\mathrm{pF},\mathrm{DP})}_{\text{flow}}\end{array}\right\}\text{ for } a_{110}>0, a_{220}<0; \tag{1.91}$$

$$\left\{\begin{array}{l}(a_{111},\dot{x}_2)\\(\dot{x}_1,a_{221})\end{array}\right\}=\left\{\begin{array}{c}\underbrace{(\mathrm{UP},\mathrm{pF})}_{\text{flow}}\\\underbrace{(\mathrm{nF},\mathrm{UP})}_{\text{flow}}\end{array}\right\}\text{ for } a_{110}<0, a_{220}<0. \tag{1.92}$$

The corresponding phase portraits ($a_{111} < a_{221}$ and $\alpha_{11} > 0$, $\alpha_{12} < 0$) for two-dimensional systems on the x_1-direction with $y_1^* = a_{111}$ and on the x_2-direction with $y_1^* = a_{221}$ are presented in Fig. 1.7a–d. The one-dimensional source and sink flows for $y_1 = a_{111} - b_{111}$ are also presented.

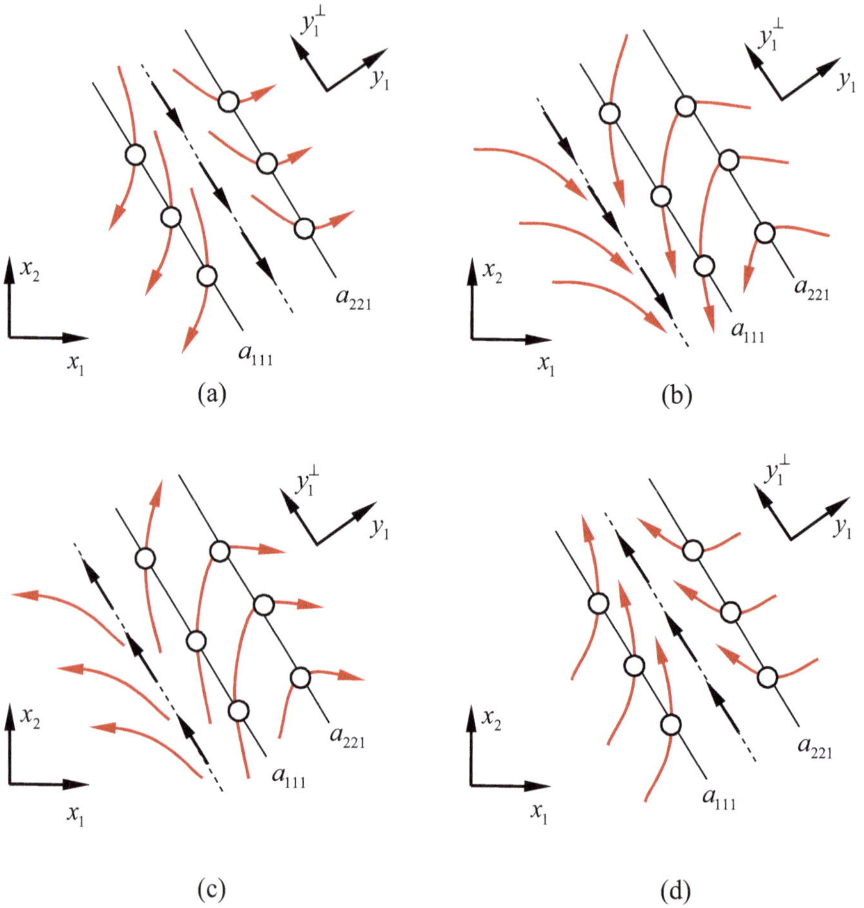

Fig. 1.6 Phase portraits ($a_{111} < a_{221}$ and $\alpha_{11} > 0$, $\alpha_{12} > 0$) for two-dimensional systems on the x_1-direction with $y_1^* = a_{111}$ and on the x_2-direction with $y_1^* = a_{221}$: (**a**) (DP,nF)-flow of $y_1^* = a_{111}$ and (pF,UP)-flow of $y_1^* = a_{221}$ ($a_{110} > 0$, $a_{220} > 0$), (**b**) (UP,nF)-flow of $y_1^* = a_{111}$ and (nF,DP)-flow of $y_1^* = a_{221}$ ($a_{110} < 0$, $a_{220} > 0$), (**c**) (UP,pF)-flow of $y_1^* = a_{111}$ and (pF,DP)-flow of $y_1^* = a_{221}$ ($a_{110} > 0$, $a_{220} < 0$), (**d**) (DP,pF)-flow of $y_1^* = a_{111}$ and (nF,UP)-flow of $y_1^* = a_{221}$ ($a_{110} < 0$, $a_{220} < 0$)

1.3.3 Inflection Source and Sink

For $a_{221} = a_{111}$, the dynamical system becomes

$$\begin{aligned} \dot{x}_2 &= a_{220}(y_1 - a_{111}), \\ \dot{x}_1 &= a_{110}(y_1 - a_{221}), \end{aligned} \tag{1.93}$$

where

$$y_1 = \alpha_{11}x_1 + \alpha_{12}x_2, \tag{1.94}$$

with the first integral manifold as

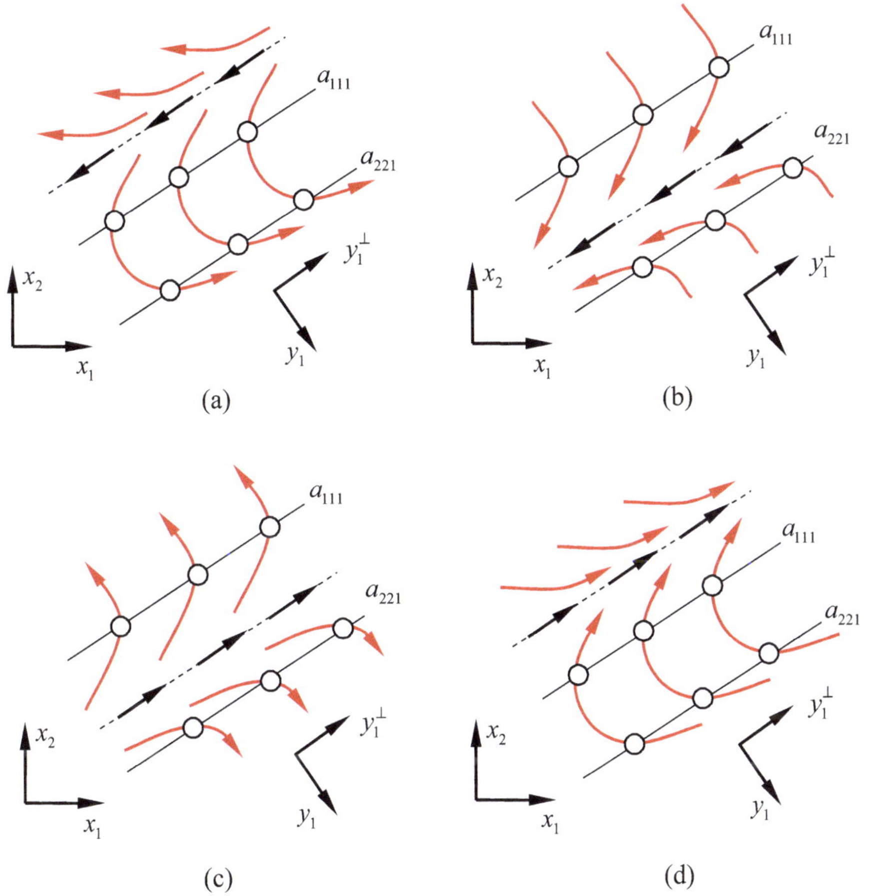

Fig. 1.7 Phase portraits ($a_{111} < a_{221}$ and $\alpha_{11} > 0$, $\alpha_{12} < 0$) for two-dimensional systems on the x_1-direction with $y_1^* = a_{111}$ and on the x_2-direction with $y_1^* = a_{221}$: (**a**) (UP,nF)-flow of $y_1^* = a_{111}$ and (pF,UP)-flow of $y_1^* = a_{221}$ ($a_{110} > 0$, $a_{220} > 0$), (**b**) (DP,nF)-flow of $y_1^* = a_{111}$ and (nF,DP)-flow of $y_1^* = a_{221}$ ($a_{110} < 0$, $a_{220} > 0$), (**c**) (DP,pF)-flow of $y_1^* = a_{111}$ and (pF,DP)-flow of $y_1^* = a_{221}$ ($a_{110} > 0$, $a_{220} < 0$), (**d**) (UP,pF)-flow of $y_1^* = a_{111}$ and (pF,UP)-flow of $y_1^* = a_{221}$ ($a_{110} < 0$, $a_{220} < 0$)

$$a_{110}(x_2 - x_{20}) = a_{220}(x_1 - x_{10}). \tag{1.95}$$

A dynamical system in the y_1-direction is

$$\dot{y}_1 = b_{110}(y_1 - a_{111}); \tag{1.96}$$

and a dynamical system in the $y_1^\perp$-direction is

$$\dot{y}_1^\perp = c_{110}(y_1 - a_{111}). \tag{1.97}$$

The flow along the y_1 -direction is given by

$$t - t_0 = \frac{1}{b_{110}} \ln \frac{|y_1 - a_{111}|}{|y_{10} - a_{111}|}. \tag{1.98}$$

The two-dimensional dynamical system possesses the infinite-equilibriums for $\alpha_{11} > 0$ and $\alpha_{12} > 0$ as

$$
\begin{aligned}
(a_{111}, a_{221}) &= \underbrace{(\mathrm{II}, \mathrm{SO})}_{\text{increasing-inflection source}} \quad \text{for } a_{110} > 0, a_{220} > 0, G_{110} > 0;\\
(a_{111}, a_{221}) &= \underbrace{(\mathrm{DI}, \mathrm{SO})}_{\text{decreasing-inflection source}} \quad \text{for } a_{110} < 0, a_{220} > 0, G_{110} > 0,\\
(a_{111}, a_{221}) &= \underbrace{(\mathrm{DI}, \mathrm{SI})}_{\text{decreasing-inflection sink}} \quad \text{for } a_{110} < 0, a_{220} > 0, G_{110} < 0;\\
(a_{111}, a_{221}) &= \underbrace{(\mathrm{DI}, \mathrm{SO})}_{\text{decreasing-inflection source}} \quad \text{for } a_{110} > 0, a_{220} < 0, G_{110} > 0,\\
(a_{111}, a_{221}) &= \underbrace{(\mathrm{DI}, \mathrm{SI})}_{\text{decreasing-inflection sink}} \quad \text{for } a_{110} > 0, a_{220} < 0, G_{110} > 0;\\
(a_{111}, a_{221}) &= \underbrace{(\mathrm{II}, \mathrm{SI})}_{\text{increasing-inflection source}} \quad \text{for } a_{110} < 0, a_{220} < 0, G_{110} < 0.
\end{aligned}
\tag{1.99}
$$

which are presented in Fig. 1.8a–f, respectively.

The two-dimensional dynamical system possesses the infinite-equilibriums for $\alpha_{11} > 0$ and $\alpha_{12} < 0$ as

$$
\begin{aligned}
(a_{111}, a_{221}) &= \underbrace{(\mathrm{II}, \mathrm{SO})}_{\text{increasing-inflection source}} \quad \text{for } a_{110} > 0, a_{220} > 0, G_{110} > 0,\\
(a_{111}, a_{221}) &= \underbrace{(\mathrm{II}, \mathrm{SI})}_{\text{increasing-inflection sink}} \quad \text{for } a_{110} > 0, a_{220} > 0, G_{110} < 0;\\
(a_{111}, a_{221}) &= \underbrace{(\mathrm{DI}, \mathrm{SI})}_{\text{decreasing-inflection sink}} \quad \text{for } a_{110} < 0, a_{220} > 0, G_{110} < 0;\\
(a_{111}, a_{221}) &= \underbrace{(\mathrm{DI}, \mathrm{SO})}_{\text{decreasing-inflection source}} \quad \text{for } a_{110} > 0, a_{220} < 0, G_{110} > 0;\\
(a_{111}, a_{221}) &= \underbrace{(\mathrm{II}, \mathrm{SO})}_{\text{increasing-inflection source}} \quad \text{for } a_{110} < 0, a_{220} < 0, G_{110} < 0,\\
(a_{111}, a_{221}) &= \underbrace{(\mathrm{II}, \mathrm{SI})}_{\text{increasing-inflection sink}} \quad \text{for } a_{110} < 0, a_{220} < 0, G_{110} > 0.
\end{aligned}
\tag{1.100}
$$

which are presented in Fig. 1.9a–f, respectively.

Reference

A.C.J. Luo, *Two-dimensional, Quadratic Nonlinear Systems, Volume II: Bivariate Vector Fields* (HEP/Springer, Beijing/Heidelberg, 2022)

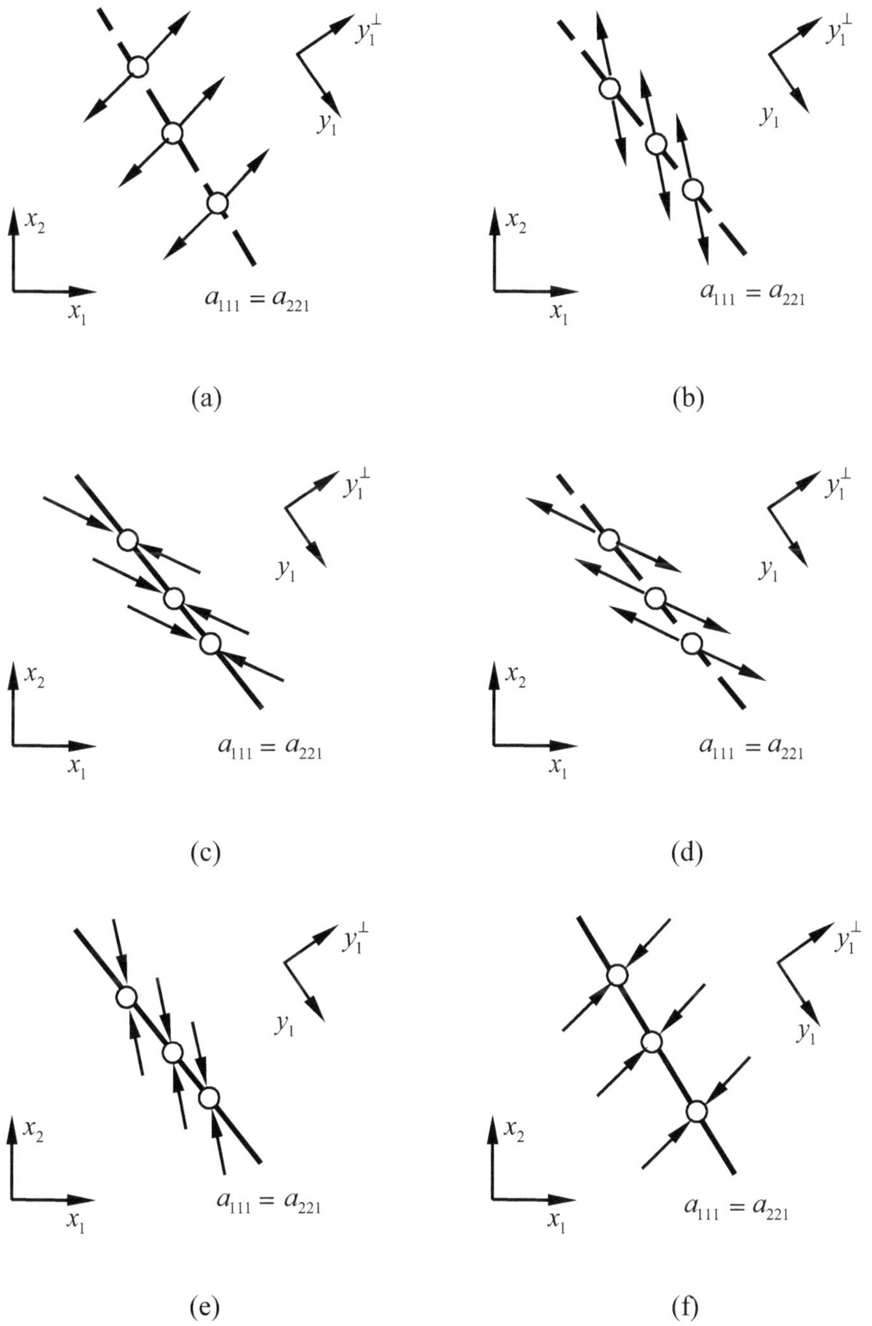

Fig. 1.8 Phase portraits for two-dimensional systems with $y_1^* = a_{111} = a_{211}$: (**a**, **b**) ($a_{110} > 0$, $a_{220} > 0$), (**c**) ($a_{110} < 0$, $a_{220} > 0$), (**d**) a ($a_{110} > 0$, $a_{220} < 0$), (**e**, **f**) ($a_{110} < 0$, $a_{220} < 0$). ($\alpha_{11} > 0$, $\alpha_{12} > 0$)

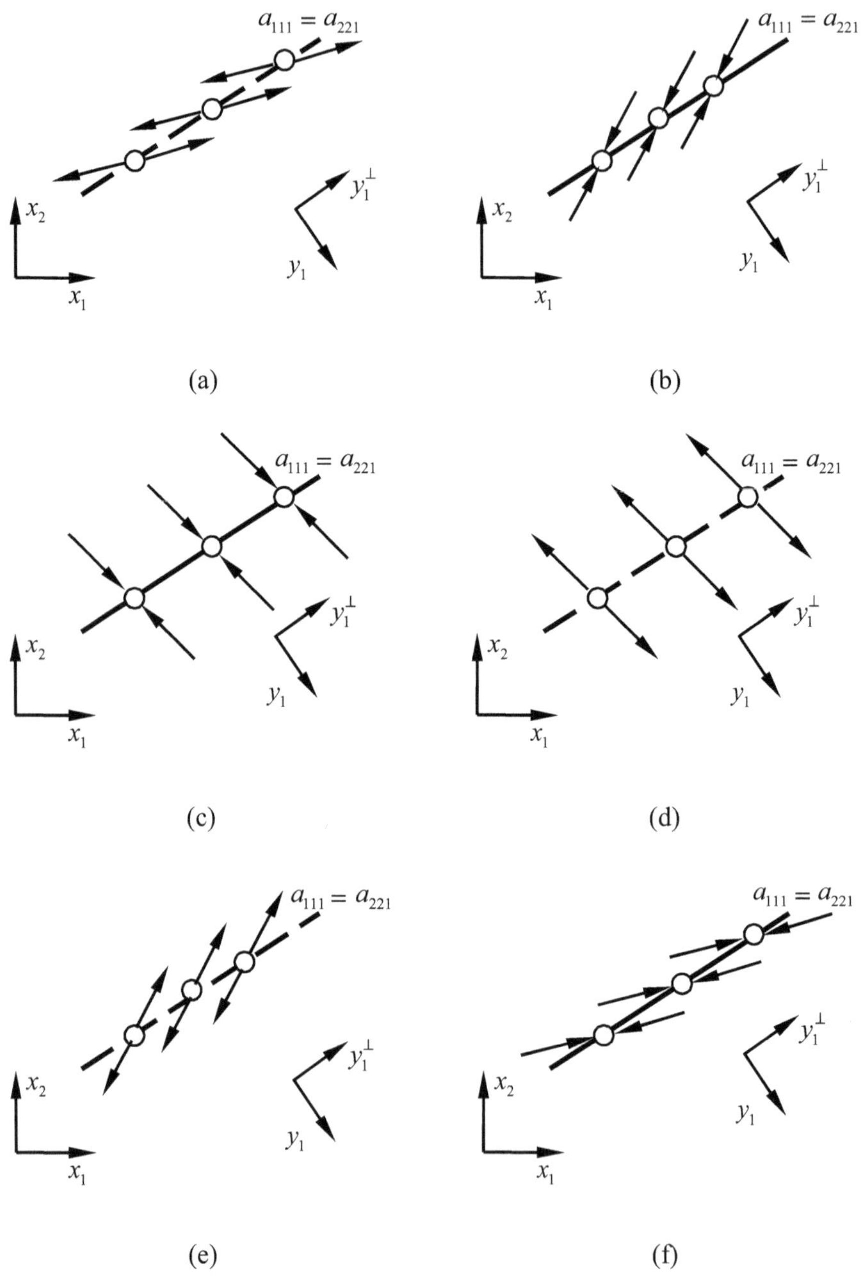

Fig. 1.9 Phase portraits for two-dimensional systems with $y_1^* = a_{111} = a_{211}$: (**a**, **b**) ($a_{110} > 0$, $a_{220} > 0$), (**c**) ($a_{110} < 0$, $a_{220} > 0$), (**d**) a ($a_{110} > 0$, $a_{220} < 0$), (**e**, **f**) ($a_{110} < 0$, $a_{220} < 0$). ($\alpha_{11} > 0$, $\alpha_{12} < 0$)

Chapter 2
Constant and Linear-Bivariate Quadratic Systems

In this chapter, a quadratic dynamical system possessing constant and linear-bivariate quadratic vector fields will be discussed, and the corresponding inflection flows exist for the appearing bifurcation of up- and down-parabola flows.

2.1 Inflection and Parabola Flows

For a dynamical system with constant and linear-bivariate quadratic vector field, the corresponding dynamical behaviors and bifurcations will be presented through the following theorem.

Theorem 2.1 *Consider a two-dimensional, nonlinear dynamical system as*

$$\begin{aligned}\dot{x}_{j_1} &= f_{j_1}(\mathbf{x},\mathbf{p}) = A_{j_1j_12} + A_{j_1j_11}y_{j_1} + A_{j_1j_10}y_{j_1}^2,\\ \dot{x}_{j_2} &= f_{j_2}(\mathbf{x},\mathbf{p}) = A_{j_2j_22} + A_{j_2j_21}y_{j_1} + A_{j_2j_20}y_{j_1}^2,\end{aligned} \tag{2.1}$$

with

$$\begin{aligned}y_{j_1} &= \alpha_{j_1j_1}x_{j_1} + \alpha_{j_1j_2}x_{j_2} \text{ for } j_1,j_2 \in \{1,2\}, j_1 \neq j_2;\\ \Delta_{jj} &= (A_{jj1})^2 - 4A_{jj2}A_{jj0} \text{ for } j = j_1, j_2.\end{aligned} \tag{2.2}$$

(i) For $A_{j_2j_21} = A_{j_2j_20} = 0$ *(*$j_1, j_2 \in \{1,2\}, j_1 \neq j_2$*)*

$$\begin{aligned}\dot{x}_{j_2} &= A_{j_2j_22} = 0,\\ \dot{x}_{j_1} &= A_{j_1j_12} + A_{j_1j_11}y_{j_1} + A_{j_1j_10}y_{j_1}^2.\end{aligned} \tag{2.3}$$

A. C. J. Luo, *Introduction to Infinite-Equilibriums in Dynamical Systems*,
https://doi.org/10.1007/978-3-031-89083-3_2

If $A_{j_2j_2 2}=0$, *the two-dimensional dynamical system becomes a one-dimensional quadratic nonlinear system in the* x_{j_1}*-direction.*
(i₁) For $\Delta_{j_1j_1}<0$, *the one-dimensional system becomes*

$$\begin{aligned}\dot{x}_{j_2}&=A_{j_2j_2 2}=0,\\ \dot{x}_{j_1}&=a_{j_1j_1 0}[(y_{j_1}-a_{j_1j_1 1})^2+b_{j_1j_1}];\end{aligned}\tag{2.4}$$

with

$$a_{j_1j_1 1}=-\frac{A_{j_1j_1 1}}{A_{j_1j_1 0}},a_{j_1j_1 0}=A_{j_1j_1 0}.\tag{2.5}$$

- *For* $a_{j_1j_1 0}>0$, *the flow in the* x_{j_1}*-direction is a positive flow.*
- *For* $a_{j_1j_1 0}<0$, *the flow in the* x_{j_1}*-direction is a negative flow.*

(i₂) For $\Delta_{j_1j_1}=0$, *the one-dimensional system becomes*

$$\begin{aligned}\dot{x}_{j_2}&=A_{j_2j_2 2}=0,\\ \dot{x}_{j_1}&=a_{j_1j_1 0}(y_{j_1}-a_{j_1j_1 1})^2;\end{aligned}\tag{2.6}$$

with

$$a_{j_1j_1 1}=\frac{A_{j_1j_1 1}}{2A_{j_1j_1 0}},a_{j_1j_1 0}=A_{j_1j_1 0}.\tag{2.7}$$

- *For* $\alpha_{j_1j_1}a_{j_1j_1 0}>0$, *the line of* $y_{j_1}=a_{j_1j_1 1}$ *in the* x_{j_1}*-direction is an upper-saddle infinite-equilibrium.*
- *For* $\alpha_{j_1j_1}a_{j_1j_1 0}<0$, *the line of* $y_{j_1}=a_{j_1j_1 1}$ *in the* x_{j_1}*-direction is a lower-saddle infinite-equilibrium.*

(i₃) For $\Delta_{j_1j_1}>0$, *the one-dimensional system becomes*

$$\begin{aligned}\dot{x}_{j_2}&=A_{j_2j_2 2}=0,\\ \dot{x}_{j_1}&=a_{j_1j_1 0}(y_{j_1}-a_{j_1j_1 1})(y_{j_1}-a_{j_1j_1 2});\end{aligned}\tag{2.8}$$

with

$$\begin{aligned}&\{a_{j_1j_1 1},a_{j_1j_1 2}\}=\text{sort}\{b_{j_1j_1 1},b_{j_1j_1 2}\},a_{j_1j_1 1}\le a_{j_1j_1 2},\\ &b_{j_1j_1 1},b_{j_1j_1 2}=\frac{A_{j_1j_1 1}}{2A_{j_1j_1 0}}\pm\frac{1}{2A_{j_1j_1 0}}\sqrt{\Delta_{j_1j_1}},a_{j_1j_1 0}=A_{j_1j_1 0}.\end{aligned}\tag{2.9}$$

- *For* $a_{j_1j_1 0}\alpha_{j_1j_1}>0$, *the lines of* $y^{*}_{j_1}=a_{j_1j_1 1},a_{j_1j_1 2}$ *in the* x_{j_1}*-direction are sink and source infinite-equilibriums, respectively.*

- *For $a_{j_1j_1 0}\alpha_{j_1j_1} < 0$, the lines of $y^*_{j_1} = a_{j_1j_1 1}, a_{j_1j_1 2}$ in the x_{j_1}-direction are source and sink infinite-equilibriums, respectively.*

(ii) If

$$A_{j_2j_2 2} \neq 0 \text{ and } \Delta_{j_1j_1} < 0, \tag{2.10}$$

the standard form of the dynamical system is

$$\begin{aligned} \dot{x}_{j_2} &= A_{j_2j_2 2}, \\ \dot{x}_{j_1} &= a_{j_1j_1 0}[(y_{j_1} - a_{j_1j_1 1})^2 + b_{j_1j_1}]. \end{aligned} \tag{2.11}$$

the two-dimensional dynamical system does not have any equilibriums for the x_{j_1}-direction on the x_{j_1}-direction.

- *For $a_{j_1j_1 0} > 0, A_{j_2j_2 2} > 0$, the flow is a* (pF,pF)*-flow, i.e.,*

$$(\dot{x}_{j_1}, \dot{x}_{j_2}) = \underbrace{(\text{pF}, \text{pF})}_{\text{flow}}. \tag{2.12}$$

- *For $a_{j_1j_1 0} < 0, A_{j_2j_2 2} > 0$, the flow is an* (nF,pF)*-flow, i.e.,*

$$(\dot{x}_{j_1}, \dot{x}_{j_2}) = \underbrace{(\text{nF}, \text{pF})}_{\text{flow}}. \tag{2.13}$$

- *For $a_{j_1j_1 0} > 0, A_{j_2j_2 2} > 0$, the flow is a* (pF,nF)*-flow, i.e.,*

$$(\dot{x}_{j_1}, \dot{x}_{j_2}) = \underbrace{(\text{pF}, \text{nF})}_{\text{flow}}. \tag{2.14}$$

- *For $a_{j_1j_1 0} < 0, A_{j_2j_2 2} < 0$, the flow is an* (nF,nF)*-flow, i.e.,*

$$(\dot{x}_{j_1}, \dot{x}_{j_2}) = \underbrace{(\text{nF}, \text{nF})}_{\text{flow}}. \tag{2.15}$$

(iii) If

$$A_{j_2j_2 2} \neq 0 \text{ and } \Delta_{j_1j_1} = 0, \tag{2.16}$$

the two-dimensional dynamical system has an equilibrium for the x_{j_1}-direction, i.e.,

$$y^*_{j_1} = -a_{j_1j_1 1}, a_{j_1j_1 0} = A_{j_1j_1 0} \text{ for the } x_{j_1}\text{-direction.} \tag{2.17}$$

The corresponding dynamical systems becomes

$$\begin{aligned}\dot{x}_{j_2} &= A_{j_2 j_2 2},\\ \dot{x}_{j_1} &= a_{j_1 j_1 0}(y_{j_1} - a_{j_1 j_1 1})^2.\end{aligned} \tag{2.18}$$

There is a dynamical system in the y_{j_1}-direction as

$$\dot{y}_{j_1} = \alpha_{j_1 j_2} A_{j_2 j_2 2} + \alpha_{j_1 j_1} a_{j_1 j_1 0}(y_{j_1} - a_{j_1 j_1 1})^2, \tag{2.19}$$

with a dynamical system in the $y_{j_1}^{\perp}$-direction as

$$\dot{y}_{j_1}^{\perp} = \beta_{j_1 j_2} A_{j_2 j_2 1} + \beta_{j_1 j_1} a_{j_1 j_1 0}(y_{j_1} - a_{j_1 j_1 1})^2, \tag{2.20}$$

where

$$\begin{aligned}&y_{j_1}^{\perp} = \beta_{j_1 j_1} x_{j_1} + \beta_{j_1 j_2} x_{j_2}, \mathbf{n}_{j_1}^{\perp} = (\beta_{j_1 j_1}, \beta_{j_1 j_2})^{\mathrm{T}};\\ &\mathbf{n}_{j_1} \cdot \mathbf{n}_{j_1}^{\perp} = \alpha_{j_1 j_1} \beta_{j_1 j_1} + \alpha_{j_1 j_2} \beta_{j_1 j_2} = 0.\end{aligned} \tag{2.21}$$

(iii_1) If

$$\frac{\alpha_{j_1 j_2} A_{j_2 j_2 2}}{\alpha_{j_1 j_1} a_{j_1 j_1 0}} \equiv b_{j_1 j_1 1}^2 > 0, \tag{2.22}$$

then the flow along the y_{j_1} -direction is

$$\begin{aligned}&y_{j_1} - a_{j_1 j_1 1} = b_{j_1 j_1 1} \tan\left[b_{j_1 j_1 1} a_{j_1 j_1 0} \alpha_{j_1 j_1}(t - t_0) + \theta_0\right],\\ &\theta_0 = \arctan \frac{y_{j_1} - a_{j_1 j_1 1}}{b_{j_1 j_1 1}}\end{aligned} \tag{2.23}$$

without any equilibrium in the y_{j_1}-direction.

(iii_2) If

$$-\frac{\alpha_{j_1 j_2} A_{j_2 j_2 2}}{\alpha_{j_1 j_1} a_{j_1 j_1 0}} \equiv b_{j_1 j_1 1}^2 > 0, \tag{2.24}$$

then the flow along the y_{j_1} -direction is

$$\begin{aligned}t - t_0 = &\frac{1}{2b_{j_1 j_1 1}} \frac{1}{a_{j_1 j_1 0} \alpha_{j_1 j_1}} \left(\ln \frac{|y_{j_1} - a_{j_1 j_1 1} - b_{j_1 j_1 1}|}{|y_{j_1 0} - a_{j_1 j_1 1} - b_{j_1 j_1 1}|} \right.\\ &\left. - \ln \frac{|y_{j_1} - a_{j_1 j_1 1} + b_{j_1 j_1 1}|}{|y_{j_1 0} - a_{j_1 j_1 1} + b_{j_1 j_1 1}|} \right),\end{aligned} \tag{2.25}$$

with the equilibriums of $y_1^* - a_{j_1 j_1 1} = -b_{j_1 j_1 1}, b_{j_1 j_1 1}$ in the y_{j_1}-direction.

(iii$_3$) The equilibrium on the line of $y^*_{j_1} = a_{j_1j_1 1}$ for the x_{j_1}-direction with $\dot{x}^*_{j_2} = A_{j_2j_2 2}$ possesses the following properties:

- *For $a_{j_1j_1 0} > 0$ and $A_{j_2j_2 2} > 0$,*

$$(a_{j_1j_1 1}, \dot{x}^*_{j_2}) = \underbrace{(\mathrm{II}, \mathrm{pF})}_{\text{increasing-inflection flow } (+)} . \tag{2.26}$$

- *For $a_{j_1j_1 0} < 0$ and $A_{j_2j_2 2} > 0$,*

$$(a_{j_1j_1 1}, \dot{x}^*_{j_2}) = \underbrace{(\mathrm{DI}, \mathrm{pF})}_{\text{decreasing-inflection flow } (+)} . \tag{2.27}$$

- *For $a_{j_1j_1 0} > 0$ and $A_{j_2j_2 2} < 0$,*

$$(a_{j_1j_1 1}, \dot{x}^*_{j_2}) = \underbrace{(\mathrm{DI}, \mathrm{nF})}_{\text{decreasing-inflection flow } (-)} . \tag{2.28}$$

- *For $a_{j_1j_1 0} < 0$ and $A_{j_2j_2 2} < 0$,*

$$(a_{j_1j_1 1}, \dot{x}^*_{j_2}) = \underbrace{(\mathrm{II}, \mathrm{nF})}_{\text{increasing-inflection flow } (-)} . \tag{2.29}$$

The corresponding increasing-inflection and decreasing-inflection flows are the appearing bifurcations for up-parabola flow and down-parabola flow.

(iv) If

$$A_{j_2j_2 2} \neq 0 \text{ and } \Delta_{j_1j_1} > 0, \tag{2.30}$$

the two-dimensional dynamical system has two simple equilibriums for the x_{j_1} direction, i.e.,

$$y^*_{j_1} = a_{j_1j_1 1}, a_{j_1j_1 2} \text{ for the } x_{j_1}\text{-direction}, \tag{2.31}$$

where

$$\begin{gathered} \{a_{j_1j_1 1}, a_{j_1j_1 2}\} = \mathrm{sort}\{b_{j_1j_1 1}, b_{j_1j_1 2}\}, a_{j_1j_1 1} < a_{j_1j_1 2}; \\ b_{j_1j_1 1}, b_{j_1j_1 2} = -\frac{1}{2A_{j_1j_1 0}}\left(A_{j_1j_1 1} + \sqrt{\Delta_{j_1j_1}}\right), a_{j_1j_1 0} = A_{j_1j_1 0}, \end{gathered} \tag{2.32}$$

and the corresponding dynamical system becomes

$$\begin{aligned}\dot{x}_{j_2} &= A_{j_2j_2 2},\\ \dot{x}_{j_1} &= a_{j_1j_1 0}(y_{j_1} - a_{j_1j_1 1})(y_{j_1} - a_{j_1j_1 2}).\end{aligned} \tag{2.33}$$

There is a dynamical system in the y_1*-direction as*

$$\dot{y}_{j_1} = \alpha_{j_1j_2}A_{j_2j_2 2} + \alpha_{j_1j_1}a_{j_1j_1 0}(y_{j_1} - a_{j_1j_1 1})(y_{j_1} - a_{j_1j_1 2}), \tag{2.34}$$

with a dynamical system in the $y_{j_1}^{\perp}$*-direction as*

$$\dot{y}_{j_1}^{\perp} = \beta_{j_1j_1}A_{j_2j_2 2} + \beta_{j_1j_2}a_{j_1j_1 0}(y_{j_1} - a_{j_1j_1 1})(y_{j_1} - a_{j_1j_1 2}). \tag{2.35}$$

The equilibrium of $y_{j_1}^{*} = a_{j_1j_1 s}$ $(s, l \in \{1,2\}, l \neq s)$ *for the* x_{j_1}*-direction with* $\dot{x}_{j_2}^{*} = A_{j_2j_2 2}$ *possesses the following flows*

- For $\alpha_{j_1j_2}a_{j_1j_1 0}(a_{j_1j_1 s} - a_{j_1j_1 l}) > 0, A_{j_2j_2 2} > 0$,

$$(a_{j_1j_1 s}, \dot{x}_{j_2}^{*}) = \underbrace{(\text{UP}, \text{pF})}_{\text{up-parabola flow }(+)}. \tag{2.36}$$

- For $\alpha_{j_1j_2}a_{j_1j_1 0}(a_{j_1j_1 s} - a_{j_1j_1 l}) < 0, A_{j_2j_2 2} > 0$,

$$(a_{j_1j_1 s}, \dot{x}_{j_2}^{*}) = \underbrace{(\text{DP}, \text{pF})}_{\text{down-parabola flow }(+)}. \tag{2.37}$$

- For $\alpha_{j_1j_2}a_{j_1j_1 0}(a_{j_1j_1 s} - a_{j_1j_1 l}) > 0, A_{j_2j_2 2} < 0$,

$$(a_{j_1j_1 s}, \dot{x}_{j_2}^{*}) = \underbrace{(\text{DP}, \text{nF})}_{\text{down-parabola flow }(-)}. \tag{2.38}$$

- For $\alpha_{j_1j_2}a_{j_1j_1 0}(a_{j_1j_1 s} - a_{j_1j_1 l}) < 0, A_{j_2j_2 2} < 0$,

$$(a_{j_1j_1 s}, \dot{x}_{j_2}^{*}) = \underbrace{(\text{UP}, \text{nF})}_{\text{up-parabola flow }(-)}. \tag{2.39}$$

(v) Under the following condition of

$$A_{j_2j_2 2} \neq 0 \text{ and } \Delta_{j_1j_1} = 0, \tag{2.40}$$

The following appearing bifurcations of parabola flows exist.

- *For* $a_{j_1j_1 0} > 0$ *and* $A_{j_2j_2 2} > 0$,

$$\left.\begin{aligned}
&\underbrace{(\dot{x}_{j_1}, \dot{x}_{j_2})}_{(\text{pF},\text{pF})\text{-flow}} \rightleftharpoons \underbrace{(a_{j_1j_1}1, \dot{x}_{j_2})}_{\text{appearing bifurcation}} \rightleftharpoons \left\{\begin{matrix} \underbrace{(a_{j_1j_1}1, \dot{x}_{j_2})}_{(\text{DP},\text{pF})\text{-flow}} \\ \underbrace{(a_{j_1j_1}2, \dot{x}_{j_2})}_{(\text{UP},\text{pF})\text{-flow}} \end{matrix}\right\} \text{for } a_{j_1j_2} > 0; \\
&\underbrace{(\dot{x}_{j_1}, \dot{x}_{j_2})}_{(\text{nF},\text{pF})\text{-flow}} \rightleftharpoons \underbrace{(a_{j_1j_1}1, \dot{x}_{j_2})}_{\text{appearing bifurcation}} \rightleftharpoons \left\{\begin{matrix} \underbrace{(a_{j_1j_1}1, \dot{x}_{j_2})}_{(\text{UP},\text{pF})\text{-flow}} \\ \underbrace{(a_{j_1j_1}2, \dot{x}_{j_2})}_{(\text{DP},\text{pF})\text{-flow}} \end{matrix}\right\} \text{for } a_{j_1j_2} < 0.
\end{aligned}\right. \tag{2.41}$$

- *For* $a_{j_1j_1 0} < 0$ *and* $A_{j_2j_2 2} > 0$,

$$\left.\begin{aligned}
&\underbrace{(\dot{x}_{j_1}, \dot{x}_{j_2})}_{(\text{nF},\text{pF})\text{-flow}} \rightleftharpoons \underbrace{(a_{j_1j_1}1, \dot{x}_{j_2})}_{\text{appearing bifurcation}} \rightleftharpoons \left\{\begin{matrix} \underbrace{(a_{j_1j_1}1, \dot{x}_{j_2})}_{(\text{UP},\text{pF})\text{-flow}} \\ \underbrace{(a_{j_1j_1}2, \dot{x}_{j_2})}_{(\text{DP},\text{pF})\text{-flow}} \end{matrix}\right\} \text{for } a_{j_1j_2} > 0; \\
&\underbrace{(\dot{x}_{j_1}, \dot{x}_{j_2})}_{(\text{nF},\text{pF})\text{-flow}} \rightleftharpoons \underbrace{(a_{j_1j_1}1, \dot{x}_{j_2})}_{\text{appearing bifurcation}} \rightleftharpoons \left\{\begin{matrix} \underbrace{(a_{j_1j_1}1, \dot{x}_{j_2})}_{(\text{DP},\text{pF})\text{-flow}} \\ \underbrace{(a_{j_1j_1}2, \dot{x}_{j_2})}_{(\text{UP},\text{pF})\text{-flow}} \end{matrix}\right\} \text{for } a_{j_1j_2} < 0.
\end{aligned}\right. \tag{2.42}$$

- *For* $a_{j_1j_1 0} > 0$ *and* $A_{j_2j_2 2} < 0$,

$$\left.\begin{aligned}
&\underbrace{(\dot{x}_{j_1}, \dot{x}_{j_2})}_{(\text{pF},\text{nF})\text{-flow}} \rightleftharpoons \underbrace{(a_{j_1j_1}1, \dot{x}_{j_2})}_{\text{appearing bifurcation}} \rightleftharpoons \left\{\begin{matrix} \underbrace{(a_{j_1j_1}1, \dot{x}_{j_2})}_{(\text{UP},\text{pF})\text{-flow}} \\ \underbrace{(a_{j_1j_1}2, \dot{x}_{j_2})}_{(\text{DP},\text{pF})\text{-flow}} \end{matrix}\right\} \text{for } a_{j_1j_2} > 0; \\
&\underbrace{(\dot{x}_{j_1}, \dot{x}_{j_2})}_{(\text{pF},\text{nF})\text{-flow}} \rightleftharpoons \underbrace{(a_{j_1j_1}1, \dot{x}_{j_2})}_{\text{appearing bifurcation}} \rightleftharpoons \left\{\begin{matrix} \underbrace{(a_{j_1j_1}1, \dot{x}_{j_2})}_{(\text{DP},\text{pF})\text{-flow}} \\ \underbrace{(a_{j_1j_1}2, \dot{x}_{j_2})}_{(\text{UP},\text{pF})\text{-flow}} \end{matrix}\right\} \text{for } a_{j_1j_2} < 0.
\end{aligned}\right. \tag{2.43}$$

- *For* $a_{j_1j_1 0} < 0$ *and* $A_{j_2j_2 2} < 0$,

$$\begin{aligned}
&\underbrace{(\dot{x}_{j_1},\dot{x}_{j_2})}_{\text{(pF, pF)-flow}} \rightleftharpoons \underbrace{(a_{j_1j_1}1,\dot{x}_{j_2})}_{\text{appearing bifurcation}} \rightleftharpoons \left\{\begin{array}{c}\underbrace{(a_{j_1j_1}1,\dot{x}_{j_2})}_{\text{(DP, pF)-flow}}\\ \underbrace{(a_{j_1j_1}2,\dot{x}_{j_2})}_{\text{(UP, pF)-flow}}\end{array}\right\}\text{for } a_{j_1j_2}>0;\\
&\underbrace{(\dot{x}_{j_1},\dot{x}_{j_2})}_{\text{(pF, pF)-flow}} \rightleftharpoons \underbrace{(a_{j_1j_1}1,\dot{x}_{j_2})}_{\text{appearing bifurcation}} \rightleftharpoons \left\{\begin{array}{c}\underbrace{(a_{j_1j_1}1,\dot{x}_{j_2})}_{\text{(UP, pF)-flow}}\\ \underbrace{(a_{j_1j_1}2,\dot{x}_{j_2})}_{\text{(DP, pF)-flow}}\end{array}\right\}\text{for } a_{j_1j_2}<0.
\end{aligned} \tag{2.44}$$

(vi) Consider

$$\begin{aligned}
\dot{x}_{j_2} &= A_{j_2j_2}2,\\
\dot{x}_{j_1} &= a_{j_1j_1}0(x_{j_1}-a_{j_1j_1}1)(x_{j_1}-a_{j_1j_1}2),
\end{aligned} \tag{2.45}$$

with

$$a^-_{j_1j_1}1 \rightleftharpoons a^+_{j_1j_1}2 \text{ and } a^-_{j_1j_1}2 \rightleftharpoons a^+_{j_1j_1}1. \tag{2.46}$$

Under

$$A_{j_2j_2}1 \neq 0, \text{and } \Delta_{j_1j_1} = (a_{j_1j_1}1 - a_{j_1j_1}2)^2 = 0, \tag{2.47}$$

the following switching bifurcations of saddle flows exist.

- *For $a_{j_1j_1}0>0$ and $A_{j_2j_2}2>0$,*

$$\begin{aligned}
&\left\{\begin{array}{c}\underbrace{(a^-_{j_1j_1}1,\dot{x}_{j_2})}_{\text{(DP, pF)-flow}}\\ \underbrace{(a^-_{j_1j_1}2,\dot{x}_{j_2})}_{\text{(UP, pF)-flow}}\end{array}\right\} \rightleftharpoons \underbrace{(a_{j_1j_1}1,\dot{x}_{j_2})}_{\substack{\text{(II, pF)-flow}\\ \text{switching bifurcation}}} \rightleftharpoons \left\{\begin{array}{c}\underbrace{(a^+_{j_1j_1}1,\dot{x}_{j_2})}_{\text{(DP, pF)-flow}}\\ \underbrace{(a^+_{j_1j_1}2,\dot{x}_{j_2})}_{\text{(UP, pF)-flow}}\end{array}\right\}\text{for } \alpha_{j_1j_2}>0;\\
&\left\{\begin{array}{c}\underbrace{(a^-_{j_1j_1}1,\dot{x}_{j_2})}_{\text{(UP, pF)-flow}}\\ \underbrace{(a^-_{j_1j_1}2,\dot{x}_{j_2})}_{\text{(DP, pF)-flow}}\end{array}\right\} \rightleftharpoons \underbrace{(a_{j_1j_1}1,\dot{x}_{j_2})}_{\substack{\text{(II, pF)-flow}\\ \text{switching bifurcation}}} \rightleftharpoons \left\{\begin{array}{c}\underbrace{(a^+_{j_1j_1}1,\dot{x}_{j_2})}_{\text{(UP, pF)-flow}}\\ \underbrace{(a^+_{j_1j_1}2,\dot{x}_{j_2})}_{\text{(DP, pF)-flow}}\end{array}\right\}\text{for } \alpha_{j_1j_2}<0.
\end{aligned} \tag{2.48}$$

- *For* $a_{j_1j_10}<0$ *and* $A_{j_2j_22}>0$,

$$
\begin{array}{l}
\left\{\begin{array}{c}
\underbrace{(a^{-}_{j_1j_11},\dot{x}_{j_2})}_{\text{(UP, pF)-flow}}\\
\underbrace{(a^{-}_{j_1j_12},\dot{x}_{j_2})}_{\text{(DP, pF)-flow}}
\end{array}\right\}
\rightleftharpoons
\underbrace{(a_{j_1j_11},\dot{x}_{j_2})}_{\substack{\text{(DI, pF)-flow}\\ \text{switching bifurcation}}}
\rightleftharpoons
\left\{\begin{array}{c}
\underbrace{(a^{+}_{j_1j_11},\dot{x}_{j_2})}_{\text{(UP, pF)-flow}}\\
\underbrace{(a^{+}_{j_1j_12},\dot{x}_{j_2})}_{\text{(DP, pF)-flow}}
\end{array}\right\}\text{for } \alpha_{j_1j_2}>0;\\
\left\{\begin{array}{c}
\underbrace{(a^{-}_{j_1j_11},\dot{x}_{j_2})}_{\text{(DP, pF)-flow}}\\
\underbrace{(a^{-}_{j_1j_12},\dot{x}_{j_2})}_{\text{(UP, pF)-flow}}
\end{array}\right\}
\rightleftharpoons
\underbrace{(a_{j_1j_11},\dot{x}_{j_2})}_{\substack{\text{(DI, pF)-flow}\\ \text{switching bifurcation}}}
\rightleftharpoons
\left\{\begin{array}{c}
\underbrace{(a^{+}_{j_1j_11},\dot{x}_{j_2})}_{\text{(DP, pF)-flow}}\\
\underbrace{(a^{+}_{j_1j_12},\dot{x}_{j_2})}_{\text{(UP, pF)-flow}}
\end{array}\right\}\text{for } \alpha_{j_1j_2}<0.
\end{array}
\tag{2.49}
$$

- *For* $a_{j_1j_10}>0$ *and* $A_{j_2j_22}<0$,

$$
\begin{array}{l}
\left\{\begin{array}{c}
\underbrace{(a^{-}_{j_1j_11},\dot{x}_{j_2})}_{\text{(UP, nF)-flow}}\\
\underbrace{(a^{-}_{j_1j_12},\dot{x}_{j_2})}_{\text{(DP, nF)-flow}}
\end{array}\right\}
\rightleftharpoons
\underbrace{(a_{j_1j_11},\dot{x}_{j_2})}_{\substack{\text{(DI, nF)-flow}\\ \text{switching bifurcation}}}
\rightleftharpoons
\left\{\begin{array}{c}
\underbrace{(a^{+}_{j_1j_11},\dot{x}_{j_2})}_{\text{(UP, nF)-flow}}\\
\underbrace{(a^{+}_{j_1j_12},\dot{x}_{j_2})}_{\text{(DP, nF)-flow}}
\end{array}\right\}\text{for } \alpha_{j_1j_2}>0;\\
\left\{\begin{array}{c}
\underbrace{(a^{-}_{j_1j_11},\dot{x}_{j_2})}_{\text{(DP, nF)-flow}}\\
\underbrace{(a^{-}_{j_1j_12},\dot{x}_{j_2})}_{\text{(UP, nF)-flow}}
\end{array}\right\}
\rightleftharpoons
\underbrace{(a_{j_1j_11},\dot{x}_{j_2})}_{\substack{\text{(DI, nF)-flow}\\ \text{switching bifurcation}}}
\rightleftharpoons
\left\{\begin{array}{c}
\underbrace{(a^{+}_{j_1j_11},\dot{x}_{j_2})}_{\text{(DP, nF)-flow}}\\
\underbrace{(a^{+}_{j_1j_12},\dot{x}_{j_2})}_{\text{(UP, nF)-flow}}
\end{array}\right\}\text{for } \alpha_{j_1j_2}<0.
\end{array}
\tag{2.50}
$$

- *For* $a_{j_1j_10}<0$ *and* $A_{j_2j_22}<0$,

$$
\begin{array}{l}
\left\{\begin{array}{c}
\underbrace{(a^{-}_{j_1j_11},\dot{x}_{j_2})}_{\text{(DP, nF)-flow}}\\
\underbrace{(a^{-}_{j_1j_12},\dot{x}_{j_2})}_{\text{(UP, nF)-flow}}
\end{array}\right\}
\rightleftharpoons
\underbrace{(a_{j_1j_11},\dot{x}_{j_2})}_{\substack{\text{(II, nF)-flow}\\ \text{switching bifurcation}}}
\rightleftharpoons
\left\{\begin{array}{c}
\underbrace{(a^{+}_{j_1j_11},\dot{x}_{j_2})}_{\text{(DP, nF)-flow}}\\
\underbrace{(a^{+}_{j_1j_12},\dot{x}_{j_2})}_{\text{(UP, nF)-flow}}
\end{array}\right\}\text{for } \alpha_{j_1j_2}>0;\\
\left\{\begin{array}{c}
\underbrace{(a^{-}_{j_1j_11},\dot{x}_{j_2})}_{\text{(UP, nF)-flow}}\\
\underbrace{(a^{-}_{j_1j_12},\dot{x}_{j_2})}_{\text{(DP, nF)-flow}}
\end{array}\right\}
\rightleftharpoons
\underbrace{(a_{j_1j_11},\dot{x}_{j_2})}_{\substack{\text{(II, nF)-flow}\\ \text{switching bifurcation}}}
\rightleftharpoons
\left\{\begin{array}{c}
\underbrace{(a^{+}_{j_1j_11},\dot{x}_{j_2})}_{\text{(UP, nF)-flow}}\\
\underbrace{(a^{+}_{j_1j_12},\dot{x}_{j_2})}_{\text{(DP, nF)-flow}}
\end{array}\right\}\text{for } \alpha_{j_1j_2}<0.
\end{array}
\tag{2.51}
$$

2.2 Proof of Theorem 2.1

(i) and (ii), the results are straight forwarded from the differential equations.
(iii) consider

$$\begin{aligned}\dot{x}_{j_2} &= A_{j_2 j_2 2},\\ \dot{x}_{j_1} &= a_{j_1 j_1 0}(y_{j_1} - a_{j_1 j_1 1})^2.\end{aligned}$$

From the above equation, a new dynamical system in the y_{j_1}-direction is constructed by

$$\dot{y}_{j_1} = \alpha_{j_1 j_2} A_{j_2 j_2 2} + \alpha_{j_1 j_1} a_{j_1 j_1 0}(y_{j_1} - a_{j_1 j_1 1})^2,$$

and a dynamical system in the $y_{j_1}^{\perp}$-direction is constructed by

$$\dot{y}_{j_1}^{\perp} = \beta_{j_1 j_2} A_{j_2 j_2 1} + \beta_{j_1 j_1} a_{j_1 j_1 0}(y_{j_1} - a_{j_1 j_1 1})^2,$$

with

$$\begin{aligned}y_{j_1}^{\perp} &= \beta_{j_1 j_1} x_{j_1} + \beta_{j_1 j_2} x_{j_2}, \mathbf{n}_{j_1}^{\perp} = (\beta_{j_1 j_1}, \beta_{j_1 j_2})^{\mathrm{T}};\\ \mathbf{n}_{j_1} \cdot \mathbf{n}_{j_1}^{\perp} &= \alpha_{j_1 j_1} \beta_{j_1 j_1} + \alpha_{j_1 j_2} \beta_{j_1 j_2} = 0.\end{aligned}$$

(iii$_1$) If there is any equilibrium in the y_{j_1}-direction, let

$$\frac{\alpha_{j_1 j_2} A_{j_2 j_2 2}}{\alpha_{j_1 j_1} a_{j_1 j_1 0}} \equiv b_{j_1 j_1 1}^2 > 0.$$

The system on the y_{j_1}-direction gives the solution of the flow as

$$\begin{aligned}y_{j_1} - a_{j_1 j_1 1} &= b_{j_1 j_1 1} \tan\left[b_{j_1 j_1 1} a_{j_1 j_1 0} \alpha_{j_1 j_1}(t - t_0) + \theta_0\right],\\ \theta_0 &= \arctan \frac{y_{j_1} - a_{j_1 j_1 1}}{b_{j_1 j_1 1}}.\end{aligned}$$

(iii$_2$) If there is an equilibriums of $y_1^* - a_{j_1 j_1 1} = -b_{j_1 j_1 1}, b_{j_1 j_1 1}$ in the y_{j_1}-direction, let

$$-\frac{\alpha_{j_1 j_2} A_{j_2 j_2 2}}{\alpha_{j_1 j_1} a_{j_1 j_1 0}} \equiv b_{j_1 j_1 1}^2 > 0,$$

and the differential equation gives the flow along the y_{j_1} -direction as

$$t - t_0 = \frac{1}{2b_{j_1j_11}} \frac{1}{a_{j_1j_10}\alpha_{j_1j_1}} \left(\ln \frac{|y_{j_1} - a_{j_1j_11} - b_{j_1j_11}|}{|y_{j_10} - a_{j_1j_11} - b_{j_1j_11}|} - \ln \frac{|y_{j_1} - a_{j_1j_11} + b_{j_1j_11}|}{|y_{j_10} - a_{j_1j_11} + b_{j_1j_11}|} \right),$$

(iii$_3$) From the differential equation in the y_{j_1}-direction, we have

$$\frac{dx_{j_1}}{dx_{j_2}}\bigg|_{y^*_{j_1}=a_{j_1j_11}} = \frac{a_{j_1j_10}(y_{j_1} - a_{j_1j_11})^2}{A_{j_2j_22}}\bigg|_{y^*_{j_1}=a_{j_1j_11}} = 0;$$

$$\frac{d^2x_{j_1}}{dx^2_{j_2}}\bigg|_{y^*_{j_1}=a_{j_1j_11}} = \frac{2a_{j_1j_10}\alpha_{j_1j_2}(y_{j_1} - a_{j_1j_11})}{A_{j_2j_22}}\bigg|_{y^*_{j_1}=a_{j_1j_11}} = 0.$$

If

$$\frac{d^3x_{j_1}}{dx^3_{j_2}}\bigg|_{y^*_{j_1}=a_{j_1j_11}} = \frac{2a_{j_1j_10}\alpha_{j_1j_2}\alpha_{j_1j_2}}{A_{j_2j_22}} > 0,$$

then the flow on $y^*_{j_1} = a_{j_1j_11}$ is an increasing-inflection flow. If

$$\frac{d^3x_{j_1}}{dx^3_{j_2}}\bigg|_{y^*_{j_1}=a_{j_1j_11}} = \frac{2a_{j_1j_10}\alpha_{j_1j_2}\alpha_{j_1j_2}}{A_{j_2j_22}} < 0,$$

then the flow o $y^*_{j_1} = a_{j_1j_11}$ is a decreasing-inflection flow.

If $\dot{x}^*_{j_2} > 0$ and $\dot{x}^*_{j_2} < 0$, the flow on $y^*_{j_1} = a_{j_1j_11}$ in the x_{j_2}-direction is positive and negative, respectively.

Therefore, the equilibrium on the line of $y^*_{j_1} = a_{j_1j_11}$ for the x_{j_1}-direction with $x^*_{j_2} = A_{j_2j_22}$ possesses the following properties as in Eqs. (2.26)–(2.29):

- For $a_{j_1j_10} > 0$ and $A_{j_2j_22} > 0$,

$$(a_{j_1j_11}, \dot{x}^*_{j_2}) = \underbrace{(\text{II}, \text{pF})}_{\text{increasing-inflection flow } (+)}.$$

- For $a_{j_1j_10} < 0$ and $A_{j_2j_22} > 0$,

$$(a_{j_1j_11}, \dot{x}^*_{j_2}) = \underbrace{(\text{DI}, \text{pF})}_{\text{decreasing-inflection flow } (+)}.$$

- For $a_{j_1j_10}>0$ and $A_{j_2j_22}<0$,

$$(a_{j_1j_11},\dot{x}_{j_2}^{*})=\underbrace{(\text{DI},\text{nF})}_{\text{decreasing-inflection flow }(-)}.$$

- For $a_{j_1j_10}<0$ and $A_{j_2j_22}<0$,

$$(a_{j_1j_11},\dot{x}_{j_2}^{*})=\underbrace{(\text{II},\text{nF})}_{\text{increasing-inflection flow }(-)}.$$

(iv) Consider a dynamical system as

$$\begin{aligned}\dot{x}_{j_2}&=A_{j_2j_22},\\ \dot{x}_{j_1}&=a_{j_1j_10}(y_{j_1}-a_{j_1j_11})(y_{j_1}-a_{j_1j_12}).\end{aligned}$$

Similarly, a dynamical system in the y_1-direction is constructed as $\dot{y}_{j_1}=a_{j_1j_2}A_{j_2j_22}+a_{j_1j_1}a_{j_1j_10}(y_{j_1}-a_{j_1j_11})(y_{j_1}-a_{j_1j_12})$, and a dynamical system in the $y_{j_1}^{\perp}$-direction is developed as $\dot{y}_{j_1}^{\perp}=\beta_{j_1j_1}A_{j_2j_22}+\beta_{j_1j_2}a_{j_1j_10}(y_{j_1}-a_{j_1j_11})(y_{j_1}-a_{j_1j_12})$.

In phase space, consider $y_{j_1}^{*}=a_{j_1j_1s}$ ($s,\,l\in\{1,2\}$, $s\neq l$), and

$$\frac{dx_{j_1}}{dx_{j_2}}\bigg|_{y_{j_1}^{*}=a_{j_1j_1s}}=\frac{a_{j_1j_10}(y_{j_1}-a_{j_1j_1s})(y_{j_1}-a_{j_1j_1l})}{A_{j_2j_22}}\bigg|_{y_{j_1}^{*}=a_{j_1j_1s}}=0.$$

If

$$\frac{d^2x_{j_1}}{dx_{j_2}^2}\bigg|_{y_{j_1}^{*}=a_{j_1j_1s}}=\frac{a_{j_1j_10}\alpha_{j_1j_2}(a_{j_1j_1s}-a_{j_1j_1l})}{A_{j_2j_22}}>0,$$

there is an up-parabola flow in the x_{j_1}-direction. If

$$\frac{d^2x_{j_1}}{dx_{j_2}^2}\bigg|_{y_{j_1}^{*}=a_{j_1j_1s}}=\frac{a_{j_1j_10}\alpha_{j_1j_2}(a_{j_1j_1s}-a_{j_1j_1l})}{A_{j_2j_22}}<0,$$

there is a down-parabola flow in the x_{j_1}-direction.

If $\dot{x}_{j_2}^{*}>0$ and $\dot{x}_{j_2}^{*}<0$, the flow on $y_{j_1}^{*}=a_{j_1j_11}$ on the x_{j_2}-direction is positive and negative, respectively.

Therefore, the equilibrium of $y_{j_1}^{*}=a_{j_1j_1s}$ ($s,\,l\in\{1,2\}$, $l\neq s$) for the x_{j_1}-direction with $\dot{x}_{j_2}^{*}=A_{j_2j_22}$ possesses the following flows as in Eqs. (2.36)–(2.39).

- For $\alpha_{j_1j_2}a_{j_1j_10}(a_{j_1j_1s}-a_{j_1j_1l})>0, A_{j_2j_22}>0,$

$$(a_{j_1j_1s},\dot{x}_{j_2}^{*})=\underbrace{(\text{UP},\text{pF})}_{\text{up-parabola flow }(+)}.$$

- For $\alpha_{j_1j_2}a_{j_1j_10}(a_{j_1j_1s}-a_{j_1j_1l})<0, A_{j_2j_22}>0,$

$$(a_{j_1j_1s},\dot{x}_{j_2}^{*})=\underbrace{(\text{DP},\text{pF})}_{\text{down-parabola flow }(+)}.$$

- For $\alpha_{j_1j_2}a_{j_1j_10}(a_{j_1j_1s}-a_{j_1j_1l})>0, A_{j_2j_22}<0,$

$$(a_{j_1j_1s},\dot{x}_{j_2}^{*})=\underbrace{(\text{DP},\text{nF})}_{\text{down-parabola flow }(-)}.$$

- For $\alpha_{j_1j_2}a_{j_1j_10}(a_{j_1j_1s}-a_{j_1j_1l})<0, A_{j_2j_22}<0,$

$$(a_{j_1j_1s},\dot{x}_{j_2}^{*})=\underbrace{(\text{UP},\text{nF})}_{\text{up-parabola flow }(-)}.$$

(v) From $a_{j_1j_1s}=a_{j_1j_11}=a_{j_1j_1l}$, with varying parameter, there are two parallel equilibrium lines of $y_{j_1}=a_{j_1j_1s},a_{j_1j_1l}$ on the x_{j_1}- direction as in Eqs. (2.41)–(2.44).

For $a_{j_0j_00}>0, A_{j_2j_22}>0$, the increasing infection bifurcation exists. Suppose $a_{j_1j_1s}>a_{j_1j_1l}$.

- For $\alpha_{j_1j_2}>0$, the following relations exist,

$$\alpha_{j_1j_2}a_{j_1j_10}(a_{j_1j_1s}-a_{j_1j_1l})>0,$$
$$\alpha_{j_1j_2}a_{j_1j_10}(a_{j_1j_1l}-a_{j_1j_1s})<0.$$

Thus, we have

$$\underbrace{(\dot{x}_{j_1},\dot{x}_{j_2})}_{(\text{pF},\text{pF})\text{-flow}}\rightleftharpoons\underbrace{(a_{j_1j_11},\dot{x}_{j_2})}_{\substack{(\text{II},\text{pF})\text{-flow}\\ \text{appearing bifurcation}}}\rightleftharpoons\left\{\begin{array}{c}\underbrace{(a_{j_1j_11},\dot{x}_{j_2})}_{(\text{DP},\text{pF})\text{-flow}}\\ \underbrace{(a_{j_1j_12},\dot{x}_{j_2})}_{(\text{UP},\text{pF})\text{-flow}}\end{array}\right\}\text{for }\alpha_{j_1j_2}>0.$$

- For $\alpha_{j_1j_2}<0$, the following relations exist

$$\alpha_{j_1j_2}a_{j_1j_10}(a_{j_1j_1s}-a_{j_1j_1l})<0,$$
$$\alpha_{j_1j_2}a_{j_1j_10}(a_{j_1j_1l}-a_{j_1j_1s})>0.$$

Thus, we have

$$\underbrace{(\dot{x}_{j_1},\dot{x}_{j_2})}_{\text{(pF, pF)-flow}} \rightleftharpoons \underbrace{(a_{j_1j_1}1,\dot{x}_{j_2})}_{\substack{\text{(II, pF)-flow}\\ \text{appearing bifurcation}}} \rightleftharpoons \left\{\begin{array}{c}\underbrace{(a_{j_1j_1}1,\dot{x}_{j_2})}_{\text{(UP, pF)-flow}}\\ \underbrace{(a_{j_1j_1}2,\dot{x}_{j_2})}_{\text{(DP, pF)-flow}}\end{array}\right\} \text{for } \alpha_{j_1j_2}<0.$$

Therefore, for $a_{j_1j_1 0}>0$ and $A_{j_2j_2 2}>0$,

$$\underbrace{(\dot{x}_{j_1},\dot{x}_{j_2})}_{\text{(pF, pF)-flow}} \rightleftharpoons \underbrace{(a_{j_1j_1}1,\dot{x}_{j_2})}_{\substack{\text{(II, pF)-flow}\\ \text{appearing bifurcation}}} \rightleftharpoons \left\{\begin{array}{c}\underbrace{(a_{j_1j_1}1,\dot{x}_{j_2})}_{\text{(DP, pF)-flow}}\\ \underbrace{(a_{j_1j_1}2,\dot{x}_{j_2})}_{\text{(UP, pF)-flow}}\end{array}\right\} \text{for } \alpha_{j_1j_2}>0.$$

$$\underbrace{(\dot{x}_{j_1},\dot{x}_{j_2})}_{\text{(pF, pF)-flow}} \rightleftharpoons \underbrace{(a_{j_1j_1}1,\dot{x}_{j_2})}_{\substack{\text{(II, pF)-flow}\\ \text{appearing bifurcation}}} \rightleftharpoons \left\{\begin{array}{c}\underbrace{(a_{j_1j_1}1,\dot{x}_{j_2})}_{\text{(UP, pF)-flow}}\\ \underbrace{(a_{j_1j_1}2,\dot{x}_{j_2})}_{\text{(DP, pF)-flow}}\end{array}\right\} \text{for } \alpha_{j_1j_2}<0.$$

Doing the similar analysis, the following appearing bifurcations are obtained.

- For $a_{j_1j_1 0}<0$ and $A_{j_2j_2 2}>0$,

$$\underbrace{(\dot{x}_{j_1},\dot{x}_{j_2})}_{\text{(nF, pF)-flow}} \rightleftharpoons \underbrace{(a_{j_1j_1}1,\dot{x}_{j_2})}_{\substack{\text{(DI, pF)-flow}\\ \text{appearing bifurcation}}} \rightleftharpoons \left\{\begin{array}{c}\underbrace{(a_{j_1j_1}1,\dot{x}_{j_2})}_{\text{(UP, pF)-flow}}\\ \underbrace{(a_{j_1j_1}2,\dot{x}_{j_2})}_{\text{(DP, pF)-flow}}\end{array}\right\} \text{for } \alpha_{j_1j_2}>0;$$

$$\underbrace{(\dot{x}_{j_1},\dot{x}_{j_2})}_{\text{(nF, pF)-flow}} \rightleftharpoons \underbrace{(a_{j_1j_1}1,\dot{x}_{j_2})}_{\substack{\text{(DI, pF)-flow}\\ \text{appearing bifurcation}}} \rightleftharpoons \left\{\begin{array}{c}\underbrace{(a_{j_1j_1}1,\dot{x}_{j_2})}_{\text{(DP, pF)-flow}}\\ \underbrace{(a_{j_1j_1}2,\dot{x}_{j_2})}_{\text{(UP, pF)-flow}}\end{array}\right\} \text{for } \alpha_{j_1j_2}<0.$$

- For $a_{j_1j_10} > 0$ and $A_{j_2j_22} < 0$,

$$\underbrace{(\dot{x}_{j_1}, \dot{x}_{j_2})}_{\text{(pF, nF)-flow}} \rightleftharpoons \underbrace{(a_{j_1j_11}, \dot{x}_{j_2})}_{\substack{\text{(DI, nF)-flow} \\ \text{appearing bifurcation}}} \rightleftharpoons \left\{ \begin{array}{c} \underbrace{(a_{j_1j_11}, \dot{x}_{j_2})}_{\text{(UP, nF)-flow}} \\ \underbrace{(a_{j_1j_12}, \dot{x}_{j_2})}_{\text{(DP, nF)-flow}} \end{array} \right\} \text{for } \alpha_{j_1j_2} > 0;$$

$$\underbrace{(\dot{x}_{j_1}, \dot{x}_{j_2})}_{\text{(PF, nF)-flow}} \rightleftharpoons \underbrace{(a_{j_1j_11}, \dot{x}_{j_2})}_{\substack{\text{(DI, nF)-flow} \\ \text{appearing bifurcation}}} \rightleftharpoons \left\{ \begin{array}{c} \underbrace{(a_{j_1j_11}, \dot{x}_{j_2})}_{\text{(DP, nF)-flow}} \\ \underbrace{(a_{j_1j_12}, \dot{x}_{j_2})}_{\text{(UP, nF)-flow}} \end{array} \right\} \text{for } \alpha_{j_1j_2} < 0.$$

- For $a_{j_1j_10} < 0$ and $A_{j_2j_22} < 0$,

$$\underbrace{(\dot{x}_{j_1}, \dot{x}_{j_2})}_{\text{(pF, pF)-flow}} \rightleftharpoons \underbrace{(a_{j_1j_11}, \dot{x}_{j_2})}_{\substack{\text{(II, nF)-flow} \\ \text{appearing bifurcation}}} \rightleftharpoons \left\{ \begin{array}{c} \underbrace{(a_{j_1j_11}, \dot{x}_{j_2})}_{\text{(DP, nF)-flow}} \\ \underbrace{(a_{j_1j_12}, \dot{x}_{j_2})}_{\text{(UP, nF)-flow}} \end{array} \right\} \text{for } \alpha_{j_1j_2} > 0;$$

$$\underbrace{(\dot{x}_{j_1}, \dot{x}_{j_2})}_{\text{(pF, pF)-flow}} \rightleftharpoons \underbrace{(a_{j_1j_11}, \dot{x}_{j_2})}_{\substack{\text{(II, nF)-flow} \\ \text{appearing bifurcation}}} \rightleftharpoons \left\{ \begin{array}{c} \underbrace{(a_{j_1j_11}, \dot{x}_{j_2})}_{\text{(UP, nF)-flow}} \\ \underbrace{(a_{j_1j_12}, \dot{x}_{j_2})}_{\text{(DP, nF)-flow}} \end{array} \right\} \text{for } \alpha_{j_1j_2} < 0.$$

(vi) If $a_{j_1j_1s} = a_{j_1j_1l} = a_{j_1j_11}$, then

$$\begin{aligned} \frac{dx_{j_1}}{dx_{j_2}}\bigg|_{y^*_{j_1} = a_{j_1j_1s}} &= \frac{a_{j_1j_10}(y_{j_1} - a_{j_1j_1s})(y_{j_1} - a_{j_1j_1l})}{A_{j_2j_22}}\bigg|_{y^*_{j_1} = a_{j_1j_1s}} = 0, \\ \frac{d^2x_{j_1}}{dx^2_{j_2}}\bigg|_{y^*_{j_1} = a_{j_1j_1s}} &= \frac{a_{j_1j_10}\alpha_{j_1j_2}(a_{j_1j_1s} - a_{j_1j_1l})}{A_{j_2j_22}} = 0. \end{aligned}$$

Thus,

$$\frac{d^3x_{j_1}}{dx^3_{j_2}}\bigg|_{y^*_{j_1} = a_{j_1j_1s}} = \frac{2a_{j_1j_10}\alpha_{j_1j_2}\alpha_{j_1j_2}}{A_{j_2j_22}}.$$

This inflection flow is the switching bifurcation flows. If

$$\frac{d^3 x_{j_1}}{dx_{j_2}^3}\bigg|_{y_{j_1}^* = a_{j_1 j_1 1}} = \frac{2a_{j_1 j_1 0}\alpha_{j_1 j_2}\alpha_{j_1 j_2}}{A_{j_2 j_2 2}} > 0,$$

then the flow of $y_{j_1}^* = a_{j_1 j_1 1}$ is an increasing-inflection flow. If

$$\frac{d^3 x_{j_1}}{dx_{j_2}^3}\bigg|_{y_{j_1}^* = a_{j_1 j_1 1}} = \frac{2a_{j_1 j_1 0}\alpha_{j_1 j_2}\alpha_{j_1 j_2}}{A_{j_2 j_2 2}} < 0,$$

then the flow of $y_{j_1}^* = a_{j_1 j_1 1}$ is a decreasing-inflection flow.

For $a_{j_0 j_0 0} > 0, A_{j_2 j_2 2} > 0$, the increasing infection bifurcation exists. Suppose $a_{j_1 j_1 s} > a_{j_1 j_1 l}$ will be switched to $a_{j_1 j_1 s} < a_{j_1 j_1 l}$.

- For $\alpha_{j_1 j_2} > 0$, the following relations for $a_{j_1 j_1 s} > a_{j_1 j_1 l}$ exists,

$$\alpha_{j_1 j_2} a_{j_1 j_1 0}(a_{j_1 j_1 s} - a_{j_1 j_1 l}) > 0,$$
$$\alpha_{j_1 j_2} a_{j_1 j_1 0}(a_{j_1 j_1 l} - a_{j_1 j_1 s}) < 0,$$

and the following relations for $a_{j_1 j_1 s} < a_{j_1 j_1 l}$ exists,

$$\alpha_{j_1 j_2} a_{j_1 j_1 0}(a_{j_1 j_1 s} - a_{j_1 j_1 l}) < 0,$$
$$\alpha_{j_1 j_2} a_{j_1 j_1 0}(a_{j_1 j_1 l} - a_{j_1 j_1 s}) > 0.$$

So, for the before and after switching, the switching bifurcation is given by

$$\left\{\begin{matrix} \underbrace{(a_{j_1 j_1 1}^-, \dot{x}_{j_2})}_{\text{(DP, pF)-flow}} \\ \underbrace{(a_{j_1 j_1 2}^-, \dot{x}_{j_2})}_{\text{(UP, pF)-flow}} \end{matrix}\right\} \rightleftharpoons \underbrace{(a_{j_1 j_1 1}, \dot{x}_{j_2})}_{\substack{\text{(II, pF)-flow} \\ \text{switching bifurcation}}} \rightleftharpoons \left\{\begin{matrix} \underbrace{(a_{j_1 j_1 1}^+, \dot{x}_{j_2})}_{\text{(DP, pF)-flow}} \\ \underbrace{(a_{j_1 j_1 2}^+, \dot{x}_{j_2})}_{\text{(UP, pF)-flow}} \end{matrix}\right\} \text{for } \alpha_{j_1 j_2} > 0.$$

- For $\alpha_{j_1 j_2} < 0$, the following relations for $a_{j_1 j_1 s} > a_{j_1 j_1 l}$ exists,

$$\alpha_{j_1 j_2} a_{j_1 j_1 0}(a_{j_1 j_1 s} - a_{j_1 j_1 l}) < 0,$$
$$\alpha_{j_1 j_2} a_{j_1 j_1 0}(a_{j_1 j_1 l} - a_{j_1 j_1 s}) > 0,$$

and the following relations for $a_{j_1 j_1 s} < a_{j_1 j_1 l}$ exists,

$$\alpha_{j_1 j_2} a_{j_1 j_1 0}(a_{j_1 j_1 s} - a_{j_1 j_1 l}) > 0,$$
$$\alpha_{j_1 j_2} a_{j_1 j_1 0}(a_{j_1 j_1 l} - a_{j_1 j_1 s}) < 0.$$

So, for the before and after switching, the switching bifurcation is given by

$$\left\{\begin{array}{c}\underbrace{(a^{-}_{j_1j_11},\dot{x}_{j_2})}_{\text{(UP, pF)-flow}}\\ \underbrace{(a^{-}_{j_1j_12},\dot{x}_{j_2})}_{\text{(DP, pF)-flow}}\end{array}\right\} \rightleftharpoons \underbrace{(a_{j_1j_11},\dot{x}_{j_2})}_{\substack{\text{(II, pF)-flow}\\ \text{switching bifurcation}}} \rightleftharpoons \left\{\begin{array}{c}\underbrace{(a^{+}_{j_1j_11},\dot{x}_{j_2})}_{\text{(UP, pF)-flow}}\\ \underbrace{(a^{+}_{j_1j_12},\dot{x}_{j_2})}_{\text{(DP, pF)-flow}}\end{array}\right\} \text{for } \alpha_{j_1j_2}<0.$$

Therefore,

- For $a_{j_1j_10}>0$ and $A_{j_2j_22}>0$,

$$\left\{\begin{array}{c}\underbrace{(a^{-}_{j_1j_11},\dot{x}_{j_2})}_{\text{(DP, pF)-flow}}\\ \underbrace{(a^{-}_{j_1j_12},\dot{x}_{j_2})}_{\text{(UP, pF)-flow}}\end{array}\right\} \rightleftharpoons \underbrace{(a_{j_1j_11},\dot{x}_{j_2})}_{\substack{\text{(II, pF)-flow}\\ \text{switching bifurcation}}} \rightleftharpoons \left\{\begin{array}{c}\underbrace{(a^{+}_{j_1j_11},\dot{x}_{j_2})}_{\text{(DP, pF)-flow}}\\ \underbrace{(a^{+}_{j_1j_12},\dot{x}_{j_2})}_{\text{(UP, pF)-flow}}\end{array}\right\} \text{for } \alpha_{j_1j_2}>0;$$

$$\left\{\begin{array}{c}\underbrace{(a^{-}_{j_1j_11},\dot{x}_{j_2})}_{\text{(UP, pF)-flow}}\\ \underbrace{(a^{-}_{j_1j_12},\dot{x}_{j_2})}_{\text{(DP, pF)-flow}}\end{array}\right\} \rightleftharpoons \underbrace{(a_{j_1j_11},\dot{x}_{j_2})}_{\substack{\text{(II, pF)-flow}\\ \text{switching bifurcation}}} \rightleftharpoons \left\{\begin{array}{c}\underbrace{(a^{+}_{j_1j_11},\dot{x}_{j_2})}_{\text{(UP, pF)-flow}}\\ \underbrace{(a^{+}_{j_1j_12},\dot{x}_{j_2})}_{\text{(DP, pF)-flow}}\end{array}\right\} \text{for } \alpha_{j_1j_2}<0.$$

From the similar discussion, the following switching bifurcations are obtained.

- For $a_{j_1j_10}<0$ and $A_{j_2j_22}>0$,

$$\left\{\begin{array}{c}\underbrace{(a^{-}_{j_1j_11},\dot{x}_{j_2})}_{\text{(UP, pF)-flow}}\\ \underbrace{(a^{-}_{j_1j_12},\dot{x}_{j_2})}_{\text{(DP, pF)-flow}}\end{array}\right\} \rightleftharpoons \underbrace{(a_{j_1j_11},\dot{x}_{j_2})}_{\substack{\text{(DI, pF)-flow}\\ \text{switching bifurcation}}} \rightleftharpoons \left\{\begin{array}{c}\underbrace{(a^{+}_{j_1j_11},\dot{x}_{j_2})}_{\text{(UP, pF)-flow}}\\ \underbrace{(a^{+}_{j_1j_12},\dot{x}_{j_2})}_{\text{(DP, pF)-flow}}\end{array}\right\} \text{for } \alpha_{j_1j_2}>0;$$

$$\left\{\begin{array}{c}\underbrace{(a^{-}_{j_1j_11},\dot{x}_{j_2})}_{\text{(DP, pF)-flow}}\\ \underbrace{(a^{-}_{j_1j_12},\dot{x}_{j_2})}_{\text{(UP, pF)-flow}}\end{array}\right\} \rightleftharpoons \underbrace{(a_{j_1j_11},\dot{x}_{j_2})}_{\substack{\text{(DI, pF)-flow}\\ \text{switching bifurcation}}} \rightleftharpoons \left\{\begin{array}{c}\underbrace{(a^{+}_{j_1j_11},\dot{x}_{j_2})}_{\text{(DP, pF)-flow}}\\ \underbrace{(a^{+}_{j_1j_12},\dot{x}_{j_2})}_{\text{(UP, pF)-flow}}\end{array}\right\} \text{for } \alpha_{j_1j_2}<0.$$

- For $a_{j_1j_10} > 0$ and $A_{j_2j_22} < 0$,

$$\left\{\begin{array}{c}\underbrace{(a^-_{j_1j_11}, \dot{x}_{j_2})}_{\text{(UP, nF)-flow}} \\ \underbrace{(a^-_{j_1j_12}, \dot{x}_{j_2})}_{\text{(DP, nF)-flow}}\end{array}\right\} \rightleftharpoons \underbrace{(a_{j_1j_11}, \dot{x}_{j_2})}_{\substack{\text{(DI, nF)-flow} \\ \text{switching bifurcation}}} \rightleftharpoons \left\{\begin{array}{c}\underbrace{(a^+_{j_1j_11}, \dot{x}_{j_2})}_{\text{(UP, nF)-flow}} \\ \underbrace{(a^+_{j_1j_12}, \dot{x}_{j_2})}_{\text{(DP, nF)-flow}}\end{array}\right\} \text{for } \alpha_{j_1j_2} > 0;$$

$$\left\{\begin{array}{c}\underbrace{(a^-_{j_1j_11}, \dot{x}_{j_2})}_{\text{(DP, nF)-flow}} \\ \underbrace{(a^-_{j_1j_12}, \dot{x}_{j_2})}_{\text{(UP, nF)-flow}}\end{array}\right\} \rightleftharpoons \underbrace{(a_{j_1j_11}, \dot{x}_{j_2})}_{\substack{\text{(DI, nF)-flow} \\ \text{switching bifurcation}}} \rightleftharpoons \left\{\begin{array}{c}\underbrace{(a^+_{j_1j_11}, \dot{x}_{j_2})}_{\text{(DP, nF)-flow}} \\ \underbrace{(a^+_{j_1j_12}, \dot{x}_{j_2})}_{\text{(UP, nF)-flow}}\end{array}\right\} \text{for } \alpha_{j_1j_2} < 0.$$

- For $a_{j_1j_10} < 0$ and $A_{j_2j_22} < 0$,

$$\left\{\begin{array}{c}\underbrace{(a^-_{j_1j_11}, \dot{x}_{j_2})}_{\text{(DP, nF)-flow}} \\ \underbrace{(a^-_{j_1j_12}, \dot{x}_{j_2})}_{\text{(UP, nF)-flow}}\end{array}\right\} \rightleftharpoons \underbrace{(a_{j_1j_11}, \dot{x}_{j_2})}_{\substack{\text{(II, nF)-flow} \\ \text{switching bifurcation}}} \rightleftharpoons \left\{\begin{array}{c}\underbrace{(a^+_{j_1j_11}, \dot{x}_{j_2})}_{\text{(DP, nF)-flow}} \\ \underbrace{(a^+_{j_1j_12}, \dot{x}_{j_2})}_{\text{(UP, nF)-flow}}\end{array}\right\} \text{for } \alpha_{j_1j_2} > 0;$$

$$\left\{\begin{array}{c}\underbrace{(a^-_{j_1j_11}, \dot{x}_{j_2})}_{\text{(UP, nF)-flow}} \\ \underbrace{(a^-_{j_1j_12}, \dot{x}_{j_2})}_{\text{(DP, nF)-flow}}\end{array}\right\} \rightleftharpoons \underbrace{(a_{j_1j_11}, \dot{x}_{j_2})}_{\substack{\text{(II, nF)-flow} \\ \text{switching bifurcation}}} \rightleftharpoons \left\{\begin{array}{c}\underbrace{(a^+_{j_1j_11}, \dot{x}_{j_2})}_{\text{(UP, nF)-flow}} \\ \underbrace{(a^+_{j_1j_12}, \dot{x}_{j_2})}_{\text{(DP, nF)-flow}}\end{array}\right\} \text{for } \alpha_{j_1j_2} < 0.$$

The Theorem 2.1 can be proved. ■

2.3 Inflection-Flows and Parabola-Flows

In this section, singular inflection flows are presented for appearing bifurcations of up- and down-parabola flows.

2.3.1 Inflection-Flows for Parabola-Flow Appearing

Consider a dynamical system as

$$\begin{aligned}\dot{x}_2 &= A_{222}, \\ \dot{x}_1 &= a_{110}(y_1 - a_{111})^2\end{aligned} \tag{2.52}$$

with

$$y_1 = \alpha_{11}x_1 + \alpha_{12}x_2. \tag{2.53}$$

The corresponding dynamical system on the y_1-direction is

$$\dot{y}_1 = \alpha_{11}a_{110}(y_1 - a_{111})^2 + \alpha_{12}A_{222} \tag{2.54}$$

and

$$\dot{y}_1^{\perp} = \beta_{11}a_{110}(y_1 - a_{111})^2 + \beta_{12}A_{222}, \tag{2.55}$$

where

$$\begin{aligned} &y_1^{\perp} = \beta_{11}x_1 + \beta_{12}x_2, \mathbf{n}_1^{\perp} = (\beta_{11}, \beta_{12})^{\mathrm{T}}; \\ &(\mathbf{n}_1)^{\mathrm{T}} \cdot \mathbf{n}_1^{\perp} = \alpha_{11}\beta_{11} + \alpha_{12}\beta_{12} = 0. \end{aligned} \tag{2.56}$$

If

$$\frac{\alpha_{12}A_{222}}{\alpha_{11}a_{110}} \equiv b_{111}^2 > 0, \tag{2.57}$$

then the flow along the y_1-direction is

$$\begin{aligned} &y_1 - a_{111} = b_{111}\tan[b_{111}a_{110}\alpha_{11}(t - t_0) + \theta_0], \\ &\theta_0 = \arctan\frac{y_1 - a_{111}}{b_{111}} \end{aligned} \tag{2.58}$$

without any equilibrium in the y_1-direction. If

$$-\frac{\alpha_{12}A_{222}}{\alpha_{11}a_{110}} \equiv b_{111}^2 > 0, \tag{2.59}$$

then the flow along the y_1-direction is

$$t - t_0 = \frac{1}{2b_{111}}\frac{1}{a_{110}a_{11}}\left(\ln\frac{|y_1 - a_{111} - b_{111}|}{|y_{10} - a_{111} - b_{111}|} - \ln\frac{|y_1 - a_{111} + b_{111}|}{|y_{10} - a_{111} + b_{111}|}\right), \tag{2.60}$$

with the equilibriums of $y_1^* - a_{111} = -b_{111}, b_{111}$ in the y_1-direction.

For $\alpha_{11} > 0$ and $\alpha_{12} < 0$, the phase portraits on the line of $y_1 = a_{111}$ for the four cases of increasing- and decreasing-inflection flows for the x_1-direction are presented in Fig. 2.1a–d.

In Fig. 2.1a, the (II,pF)-positive increasing-inflection flow on the line of $y_1^* = a_{111}$ ($a_{110} > 0$, $A_{222} > 0$) is presented. In the x_1-direction, the saddle equilibrium on the

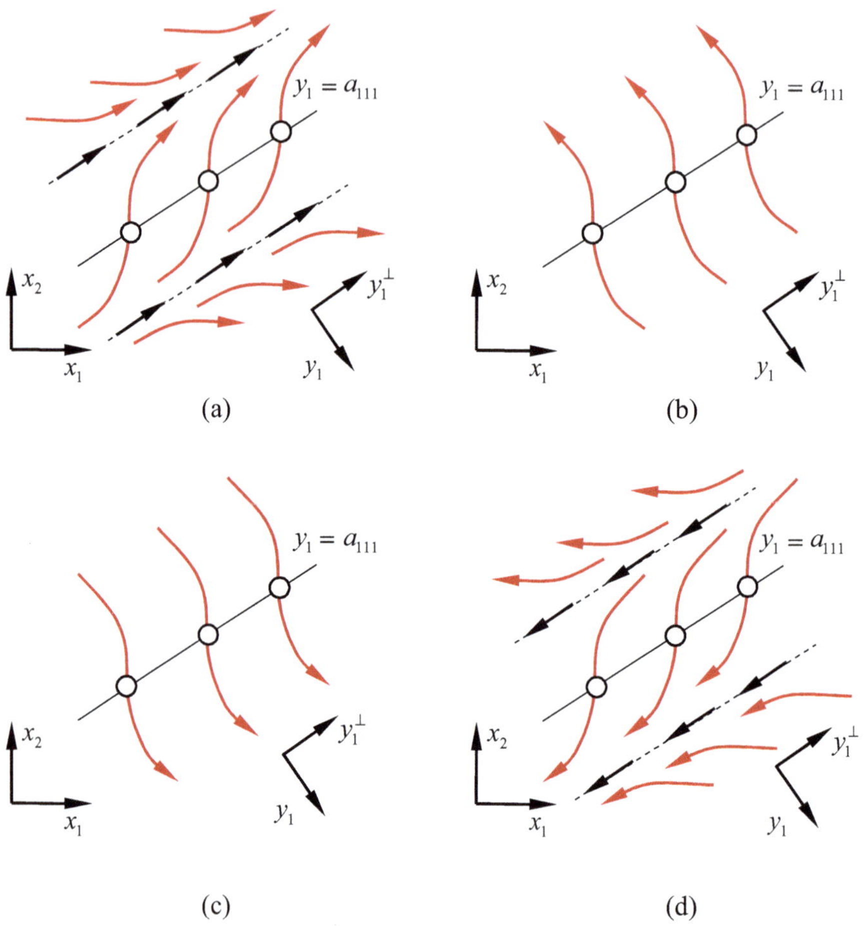

Fig. 2.1 Phase portraits for two-dimensional systems on the x_1-direction with $y_1^* = a_{111}$ and on the x_2-direction with $\dot{x}_2 = A_{222} \neq 0$ ($\alpha_{11} > 0$, $\alpha_{12} < 0$): (**a**) (II,pF) positive increasing-inflection flow on the line of $y_1^* = a_{111}$ ($a_{110} > 0$, $A_{222} > 0$), (**b**) (DI,pF) positive deceasing-inflection flow on the line of $y_1^* = a_{111}$ ($a_{110} < 0$, $A_{222} > 0$), (**c**) (DI,nF) negative increasing-inflection flow on the line of $y_1^* = a_{111}$ ($a_{110} > 0$, $A_{222} < 0$), (**d**) (II,nF) negative decreasing-inflection flow on the line of $y_1^* = a_{111}$ ($a_{110} < 0$, $A_{222} < 0$)

line of $y_1^* = a_{111}$ is an upper-saddle, and in the x_2-direction, the positive constant speed on the line of $y_1^* = a_{111}$ exist. Due to $\alpha_{12}A_{222}\alpha_{11}a_{110} < 0$, there are two equilibriums of $y_1^* - a_{111} = -b_{111}, b_{111}$ in the y_1-direction. Let $\beta_{12} = \alpha_{11} > 0$ and $\beta_{11} = -\alpha_{12} > 0$. Thus $\dot{y}_1^\perp = -\alpha_{12}a_{110}(y_1 - a_{111})^2 + \alpha_{11}A_{222} > 0$. There is a flow along the positive $y_1^\perp$ direction at the two equilibriums. Due to $\alpha_{11}a_{110} > 0$, the two flows for $y_1^* - a_{111} = -b_{111}, b_{111}$ are positive sink and source flows, respectively.

In Fig. 2.1b, the (DI,pF) positive decreasing-inflection on the line of $y_1^* = a_{111}$ ($a_{110} < 0$, $A_{222} > 0$) is presented. In the x_1-direction, the saddle equilibrium on the line of $y_1^* = a_{111}$ is a lower-saddle, and in the x_2-direction, the positive constant speed on the line of $y_1^* = a_{111}$ exist. Due to $\alpha_{12}A_{222}\alpha_{11}a_{110} > 0$, there is no equilibriums in the y_1-direction.

Similarly, the (DI,nF) negative decreasing-inflection flow on the line of $y_1^* = a_{111}$ ($a_{110} > 0, A_{222} < 0$) is presented in Fig. 2.1c. In the x_1-direction, the equilibrium on the line of $y_1^* = a_{111}$ is an upper-saddle, and in the x_2-direction, the constant speed on the line of $y_1^* = a_{111}$ is negative. Due to $\alpha_{12}A_{222}\alpha_{11}a_{110} > 0$, there is no equilibriums in the y_1-direction.

In Fig. 2.1d, the (II,nF) negative increasing-inflection flow on the line of $y_1^* = a_{111}$ ($a_{110} < 0, A_{222} < 0$) is parented. In the x_1-direction, the equilibrium on the line of $y_1^* = a_{111}$ is a lower-saddle, and in the x_2-direction, the constant speed on the line of $y_1^* = a_{111}$ is negative. Due to $\alpha_{12}A_{222}\alpha_{11}a_{110} < 0$, there are two equilibriums of $y_1^* - a_{111} = -b_{111}, b_{111}$ in the y_1-direction. Let $\beta_{12} = \alpha_{11} > 0$ and $\beta_{11} = -\alpha_{12} > 0$. Thus $\dot{y}_1^{\perp} = -\alpha_{12}a_{110}(y_1 - a_{111})^2 + \alpha_{11}A_{222} < 0$. There is a flow along the negative $y_1^{\perp}$ direction at the two equilibriums. Due to $\alpha_{11}a_{110} < 0$, the two flows for $y_1^* - a_{111} = -b_{111}, b_{111}$ are negative source and sink flows, respectively.

For $\alpha_{11} > 0$ and $\alpha_{12} > 0$, the phase portraits on the line of $y_1 = a_{111}$ for the four cases of increasing- and decreasing-inflection flows for the x_1-direction are presented in Fig. 2.2a–d.

In Fig. 2.2a, the (II,pF)-positive increasing-inflection flow on the line of $y_1^* = a_{111}$ ($a_{110} > 0, A_{222} > 0$) is presented. In the x_1-direction, the saddle equilibrium on the line of $y_1^* = a_{111}$ is an upper-saddle, and in the x_2-direction, the positive constant speed on the line of $y_1^* = a_{111}$ exist. Due to $\alpha_{12}A_{222}\alpha_{11}a_{110} > 0$, there is no equilibriums in the y_1-direction.

In Fig. 2.2b, the (DI,pF) positive decreasing-inflection flow on the line of $y_1^* = a_{111}$ ($a_{110} < 0, A_{222} > 0$) is presented. In the x_1-direction, the saddle equilibrium on the line of $y_1^* = a_{111}$ is a lower-saddle, and in the x_2-direction, the positive constant speed on the line of $y_1^* = a_{111}$ exist. Due to $\alpha_{12}A_{222}\alpha_{11}a_{110} < 0$, there are two equilibriums of $y_1^* - a_{111} = -b_{111}, b_{111}$ in the y_1-direction. Let $\beta_{12} = \alpha_{11} > 0$ and $\beta_{11} = -\alpha_{12} < 0$. Thus $\dot{y}_1^{\perp} = -\alpha_{12}a_{110}(y_1 - a_{111})^2 + \alpha_{11}A_{222} > 0$. There is a flow along the positive $y_1^{\perp}$ direction at the two equilibriums. Due to $\alpha_{11}a_{110} < 0$, the two flows for $y_1^* - a_{111} = -b_{111}, b_{111}$ are positive source and sink flows, respectively.

In Fig. 2.2c, the (DI,nF) negative decreasing-inflection flow on the line of $y_1^* = a_{111}$ ($a_{110} > 0, A_{222} < 0$) is parented. In the x_1-direction, the equilibrium on the line of $y_1^* = a_{111}$ is an upper-saddle, and in the x_2-direction, the constant speed on the line of $y_1^* = a_{111}$ is negative. Due to $\alpha_{12}A_{222}\alpha_{11}a_{110} < 0$, there are two equilibriums of $y_1^* - a_{111} = -b_{111}, b_{111}$ in the y_1-direction. Let $\beta_{12} = \alpha_{11} > 0$ and $\beta_{11} = -\alpha_{12} < 0$. Thus $\dot{y}_1^{\perp} = -\alpha_{12}a_{110}(y_1 - a_{111})^2 + \alpha_{11}A_{222} < 0$. There is a flow along the negative $y_1^{\perp}$ direction at the two equilibriums. Due to $\alpha_{11}a_{110} > 0$, the two flows for $y_1^* - a_{111} = -b_{111}, b_{111}$ are negative sink and source flows, respectively.

Similarly, the (II,nF)-negative increasing-inflection flow on the line of $y_1^* = a_{111}$ ($a_{110} < 0, A_{222} < 0$) is presented in Fig. 2.2d. In the x_1-direction, the equilibrium on the line of $y_1^* = a_{111}$ is an upper-saddle, and in the x_2-direction, the constant speed on the line of $y_1^* = a_{111}$ are negative. Due to $\alpha_{12}A_{222}\alpha_{11}a_{110} > 0$, there is no equilibriums in the y_1-direction.

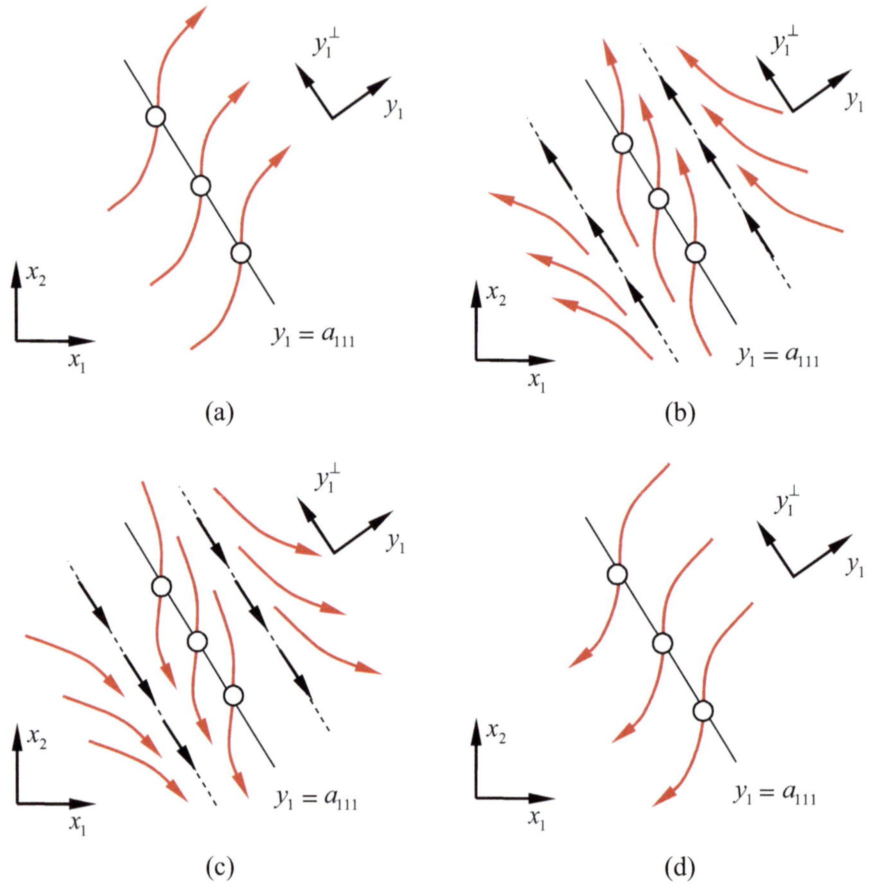

Fig. 2.2 Phase portraits for two-dimensional systems on the x_1-direction with $y_1^* = a_{111}$ and on the x_2-direction with $\dot{x}_2 = A_{222} \neq 0$ ($\alpha_{11} > 0$, $\alpha_{12} > 0$): (**a**) (II,pF) positive increasing-inflection flow on the line of $y_1^* = a_{111}$ ($a_{110} > 0$, $A_{222} > 0$), (**b**) (DI,pF) positive deceasing-inflection flow on the line of $y_1^* = a_{111}$ ($a_{110} < 0$, $A_{222} > 0$), (**c**) (DI,nF)-negative increasing-inflection flow on the line of $y_1^* = a_{111}$ ($a_{110} > 0$, $A_{222} < 0$), (**d**) (II,nF) negative decreasing-inflection flow on the line of $y_1^* = a_{111}$ ($a_{110} < 0$, $A_{222} < 0$)

2.3.2 Up- and Down-Parabola Flows

Consider a dynamical system as

$$\begin{aligned} \dot{x}_2 &= A_{222}, \\ \dot{x}_1 &= a_{110}(y_1 - a_{111})(y_1 - a_{112}), \end{aligned} \tag{2.61}$$

and

$$\dot{y}_1 = \alpha_{12}A_{222} + \alpha_{11}a_{110}(y_1 - a_{111})(y_1 - a_{112}), \tag{2.62}$$

with a dynamical system in the $y_1^{\perp}$-direction as

$$\dot{y}_1^{\perp} = \beta_{12}A_{222} + \beta_{11}a_{110}(y_1 - a_{111})(y_1 - a_{112}). \tag{2.63}$$

Let

$$c_{111} = \frac{A_{222}\alpha_{12}}{a_{110}\alpha_{11}}, \ \beta_{12} = \alpha_{11}, \beta_{11} = -\alpha_{12}. \tag{2.64}$$

If

$$(a_{111} - a_{112})^2 \geq 4c_{111}, \tag{2.65}$$

then the equilibrium for the y_1-direction is

$$y_1^* = \frac{1}{2}(a_{111} + a_{112}) \pm \frac{1}{2}\sqrt{(a_{111} - a_{112})^2 - 4c_{111}} = b_{111}, b_{112}. \tag{2.66}$$

- For $(a_{111} - a_{112})^2 = 4c_{111} > 0$, the equilibrium in the y_1-direction is

$$\begin{aligned} y_1^* &= \frac{1}{2}(a_{111} + a_{112}), \\ \dot{y}_1^{\perp} &= \alpha_{11}A_{222} + \frac{1}{4}\alpha_{12}a_{110}(a_{111} - a_{112})^2. \end{aligned} \tag{2.67}$$

Thus there exist an upper-saddle flow for $\alpha_{11}a_{110} > 0$ and a lower-saddle flow for $\alpha_{11}a_{110} > 0$ in the $y_1^{\perp}$-direction as

$$\begin{aligned} (y_1^*, \dot{y}_1^{\perp}) &= (\text{US}, \text{pF}) \ \text{ for } \ \alpha_{11}a_{110} > 0, \dot{y}_1^{\perp} > 0; \\ (y_1^*, \dot{y}_1^{\perp}) &= (\text{LS}, \text{pF}) \ \text{ for } \ \alpha_{11}a_{110} < 0, \dot{y}_1^{\perp} > 0; \\ (y_1^*, \dot{y}_1^{\perp}) &= (\text{US}, \text{nF}) \ \text{ for } \ \alpha_{11}a_{110} > 0, \dot{y}_1^{\perp} < 0; \\ (y_1^*, \dot{y}_1^{\perp}) &= (\text{LS}, \text{nF}) \ \text{ for } \ \alpha_{11}a_{110} < 0, \dot{y}_1^{\perp} < 0. \end{aligned} \tag{2.68}$$

- For $(a_{111} - a_{112})^2 > 4c_{111}$, the two equilibrium in the y_1-direction are

$$\begin{aligned} & y_{1,l}^* = b_{11l} \text{ for } l = 1, 2; \\ & \dot{y}_{1,l}^{\perp} = \alpha_{11}A_{222} + \frac{1}{4}\alpha_{12}a_{110}(b_{11l} - a_{111})(b_{11l} - a_{112}). \end{aligned} \tag{2.69}$$

Thus we have

$$
\begin{aligned}
&(y_{1,l}^{*},\dot{y}_{1,l}^{\perp})=(\mathrm{SO},\mathrm{pF})\ \text{for}\ \alpha_{11}a_{110}>0,\dot{y}_{1,l}^{\perp}>0;\\
&(y_{1,l}^{*},\dot{y}_{1,l}^{\perp})=(\mathrm{SI},\mathrm{pF})\ \text{for}\ \alpha_{11}a_{110}<0,\dot{y}_{1,l}^{\perp}>0;\\
&(y_{1,l}^{*},\dot{y}_{1,l}^{\perp})=(\mathrm{SO},\mathrm{nF})\ \text{for}\ \alpha_{11}a_{110}>0,\dot{y}_{1,l}^{\perp}<0;\\
&(y_{1,l}^{*},\dot{y}_{1,l}^{\perp})=(\mathrm{SI},\mathrm{nF})\ \text{for}\ \alpha_{11}a_{110}<0,\dot{y}_{1,l}^{\perp}<0,
\end{aligned}
\tag{2.70}
$$

which is a source flow for $\alpha_{11}a_{110} > 0$ and a sink flow for $\alpha_{11}a_{110} < 0$ in the $y_1^{\perp}$-direction.

The corresponding sink and source flows in the phase portraits are presented in Fig. 2.3. The (UP,pF)-positive up-parabola flow and (DP,pF) positive down-parabola flow on the lines of $y_1^* = a_{111}, a_{112}$ ($a_{110} > 0, A_{222} > 0$) are presented in Fig. 2.3a. For $c_{111} < 0$ due to $A_{222}\alpha_{12} < 0$, $a_{110}\alpha_{11} > 0$, the flows of $y_1^* = b_{111}, b_{112}$ exist, which are positive sink and source flows. The (DP,pF) positive down-parabola flow and (UP,pF) positive-up-parabola flow on the lines of $y_1^* = a_{111}, a_{112}$ ($a_{110} < 0, A_{222} > 0$) are presented in Fig. 2.3b. Because of $A_{222}\alpha_{12} < 0$ and $a_{110}\alpha_{11} < 0$, we have $c_{111} > 0$ but $(a_{111} - a_{112})^2 > 4c_{111}$. Thus, the positive sink and source flows of $y_1^* = b_{111}, b_{112}$ exist between $y_1^* = a_{111}, a_{112}$. In Fig. 2.3c, the (DP,nF) negative down-parabola flow and (UP,nF) negative up-parabola flow on the lines of $y_1^* = a_{111}, a_{112}$ ($a_{110} > 0$, $A_{222} < 0$) are presented. Due to $A_{222}\alpha_{12} > 0$ and $a_{110}\alpha_{11} > 0$, $c_{111} > 0$ but $(a_{111} - a_{112})^2 > 4c_{111}$. Thus, the negative sink and source flows of $y_1^* = b_{111}, b_{112}$ exist between $y_1^* = a_{111}, a_{112}$. In Fig. 2.3d, presented are the (UP,nF) negative up-parabola flow and (DP,nF) negative down-parabola flow on the lines of $y_1^* = a_{111}, a_{112}$ ($a_{110} < 0$, $A_{221} < 0$). The sink and sour flows on the lines of $y_1^* = a_{111}, a_{112}$ are the up-parabola and down-parabola flows. $c_{111} < 0$ because of $A_{222}\alpha_{12} > 0$ and $a_{110}\alpha_{11} < 0$. Thus, the negative sink and source flows of $y_1^* = b_{111}, b_{112}$ exist.

Due to $A_{222}\alpha_{12} > 0$ and $a_{110}\alpha_{11} > 0$, $c_{111} > 0$ and $(a_{111} - a_{112})^2 = 4c_{111} > 0$, the positive lower-saddle and negative upper-saddle flow of $y_1^* = b_{111} = b_{112}$ are presented in Fig. 2.4a, c. For $c_{111} > 0$ and $(a_{111} - a_{112})^2 < 4c_{111}$, the upper-saddle and lower-saddle flows of $y_1^* = b_{111}, b_{112}$ cannot exist, which are presented in Fig. 2.4b, d.

The corresponding sink and source flows in phase portraits are presented in Fig. 2.5 for $\alpha_{11} > 0$ and $\alpha_{12} > 0$. The (DP,pF) positive down-parabola flow and (UP,pF) positive up-parabola flow on the lines of $y_1^* = a_{111}, a_{112}$ ($a_{110} > 0, A_{222} > 0$) are presented in Fig. 2.5a. Because of $A_{222}\alpha_{12} > 0$ and $a_{110}\alpha_{11} > 0$, $c_{111} > 0$ but $(a_{111} - a_{112})^2 > 4c_{111}$. Thus, the positive sink and source flows of $y_1^* = b_{111}, b_{112}$ exist between $y_1^* = a_{111}, a_{112}$. The (UP,pF)-positive up-parabola flow and (DP,pF)-positive down-parabola flow on the lines of $y_1^* = a_{111}, a_{112}$ ($a_{110} < 0, A_{222} > 0$) are presented in Fig. 2.5b. $c_{111} < 0$ because $A_{222}\alpha_{12} > 0$ and $a_{110}\alpha_{11} < 0$. Thus, the flows of $y_1^* = b_{111}, b_{112}$ exist, which are positive sink and source flows. In Fig. 2.5c, the (UP,nF) negative up-parabola flow and (DP,nF) negative down-parabola flow on the lines of $y_1^* = a_{111}, a_{112}$ ($a_{110} > 0, A_{221} < 0$) are depicted. The flows on the lines of $y_1^* = a_{111}, a_{112}$ are the negative up-parabola and down-parabola flows. $c_{111} < 0$ because of $A_{222}\alpha_{12} < 0$ and $a_{110}\alpha_{11} > 0$. Thus, the negative sink and source flows of $y_1^* = b_{111}, b_{112}$ exist. In Fig. 2.5d, the (DP,nF)-negative down-parabola flow and

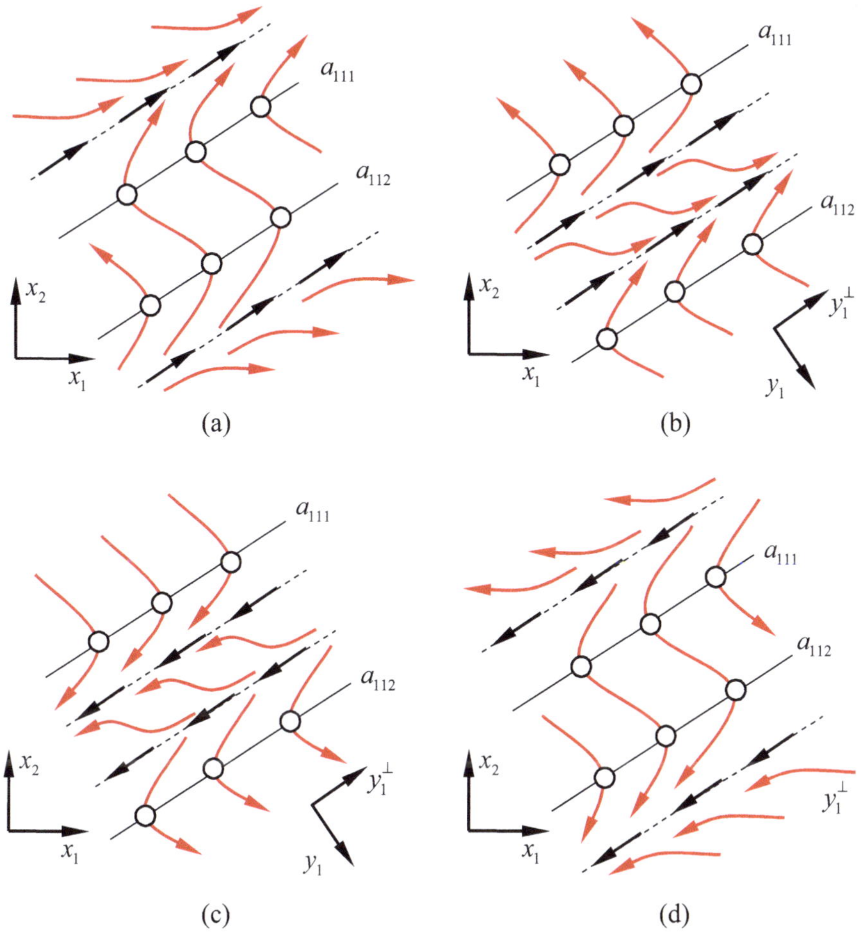

Fig. 2.3 Phase portraits for two-dimensional systems on the x_1-direction with $y_1^* = a_{111}, a_{112}$ and on the x_2-direction with $\dot{x}_2 = A_{222} \neq 0$ ($\alpha_{11} > 0$, $\alpha_{12} < 0$): (**a**) (UP,pF)-flow and (DP,pF)-flow on the lines of $y_1^* = a_{111}, a_{112}$ ($a_{110} > 0$, $A_{222} > 0$), (**b**) (DP,pF)-flow and (UP,pF)-flow on the lines of $y_1^* = a_{111}, a_{112}$ ($a_{110} < 0$, $A_{222} > 0$), (**c**) (DP,nF)-flow and (UP,nF)-flow on the lines of $y_1^* = a_{111}, a_{112}$ ($a_{110} > 0$, $A_{222} < 0$), (**d**) (UP,nF)-flow and (DP,nF)-flow on the lines of $y_1^* = a_{111}, a_{112}$ ($a_{110} < 0$, $A_{222} < 0$)

(UP,nF) negative up-parabola flow on the lines of $y_1^* = a_{111}, a_{112}$ ($a_{110} < 0, A_{222} < 0$) are presented. Due to $A_{222}\alpha_{12} < 0$ and $a_{110}\alpha_{11} < 0$, $c_{111} > 0$ but $(a_{111} - a_{112})^2 > 4c_{111}$. Thus, the negative sink and source flows of $y_1^* = b_{111}, b_{112}$ exist between $y_1^* = a_{111}, a_{112}$.

Due to $A_{222}\alpha_{12} > 0$ and $a_{110}\alpha_{11} > 0$, $c_{111} > 0$ and $(a_{111} - a_{112})^2 = 4c_{111} > 0$, the positive upper-saddle and negative lower-saddle flow of $y_1^* = b_{111} = b_{112}$ are presented in Fig. 2.6a, c. For $c_{111} > 0$ and $(a_{111} - a_{112})^2 < 4c_{111}$, the upper-saddle and lower-saddle flows of $y_1^* = b_{111}, b_{112}$ cannot exist, which are presented in Fig. 2.6b, d.

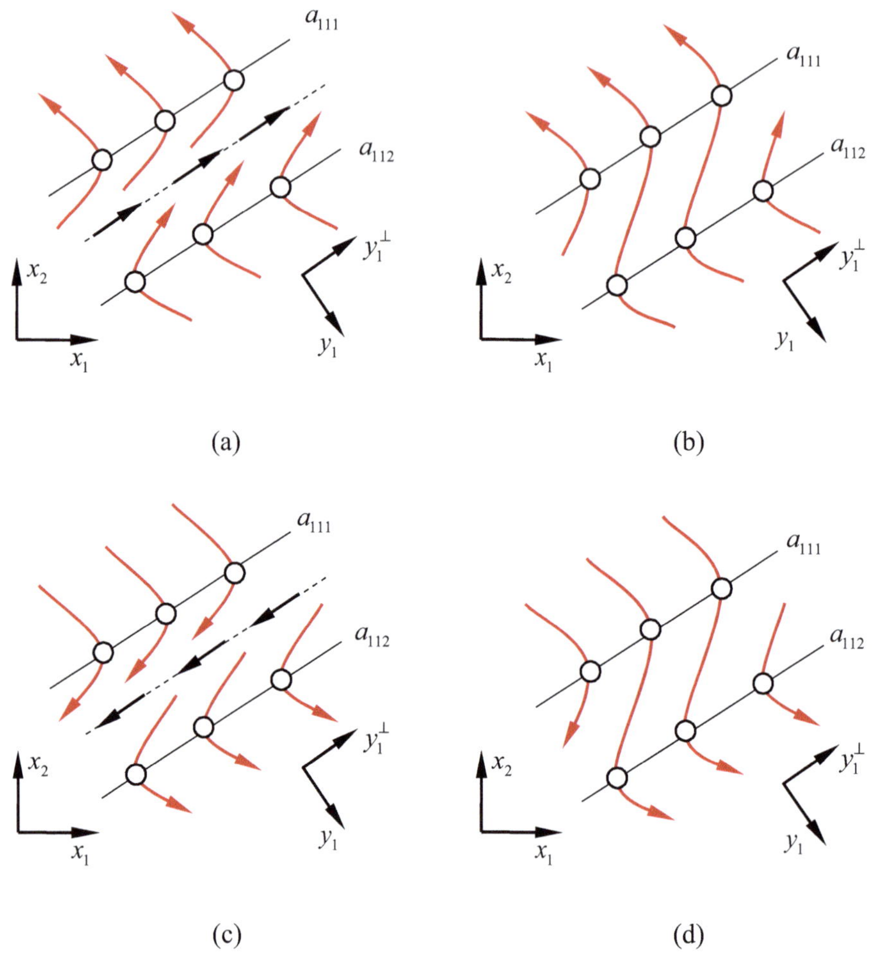

Fig. 2.4 Phase portraits for two-dimensional systems on the x_1-direction with $y_1^* = a_{111}, a_{211}$ and on the x_2-direction with $\dot{x}_2 = a_{221} \neq 0$: (**a**) a (LS,pF)-flow for (UP,pF) up-parabola flow and (DP,pF) down-parabola flow on the lines of $y_1^* = a_{111}, a_{112}$ ($a_{110} > 0, A_{222} > 0$), (**b**) without equilibrium flow for (UP,pF) up-parabola flow and (DP,pF) down-parabola flow on the lines of $y_1^* = a_{111}, a_{112}$ ($a_{110} < 0, A_{222} > 0$); (**c**) an (US,nF)-flow for (DP,nF) down-parabola flow and (UP,nF) up-parabola flow on the lines of $y_1^* = a_{111}, a_{112}$ ($a_{110} > 0$, $A_{222} < 0$), (**d**) without equilibrium flow for (DP,nF) down-parabola flow and (UP,nF) up-parabola flow on the lines of $y_1^* = a_{111}, a_{112}$ ($a_{110} < 0$, $A_{222} < 0$). ($\alpha_{11} > 0$, $\alpha_{12} < 0$)

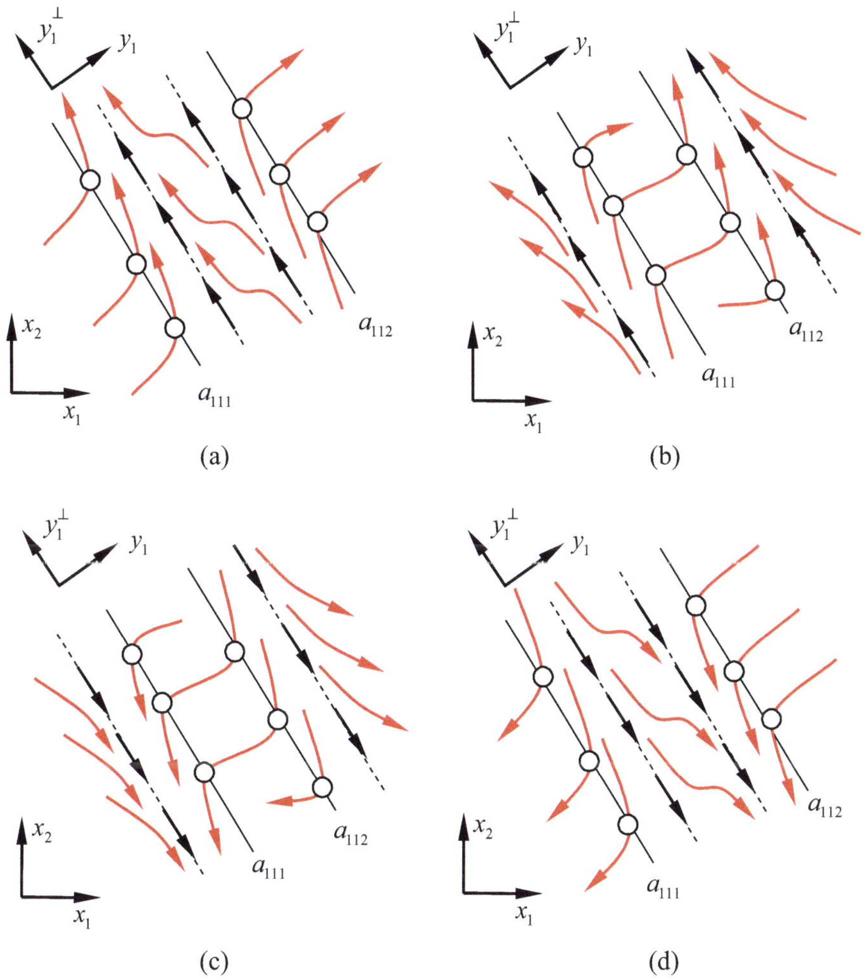

Fig. 2.5 Phase portraits for two-dimensional systems on the x_1-direction with $y_1^* = a_{111}, a_{211}$ and on the x_2-direction with $\dot{x}_2 = A_{221} \neq 0$ ($\alpha_{11} > 0$, $\alpha_{12} > 0$): (**a**) (DP,pF)-flow and (UP,pF)-flow on the lines of $y_1^* = a_{111}, a_{112}$ ($a_{110} > 0$, $A_{222} > 0$), (**b**) (UP,pF)-flow and (DP,pF)-flow on the lines of $y_1^* = a_{111}, a_{112}(a_{110} > 0, A_{222} > 0)$, (**c**) (UP,nF)-flow and (DP,nF)-flow on the lines of $y_1^* = a_{111}, a_{112}$ ($a_{110} > 0$, $A_{222} < 0$), (**d**) (DP,nF)-flow and (UP,nF)-flow on the lines of $y_1^* = a_{111}, a_{112}(a_{110} < 0$, $A_{222} < 0)$

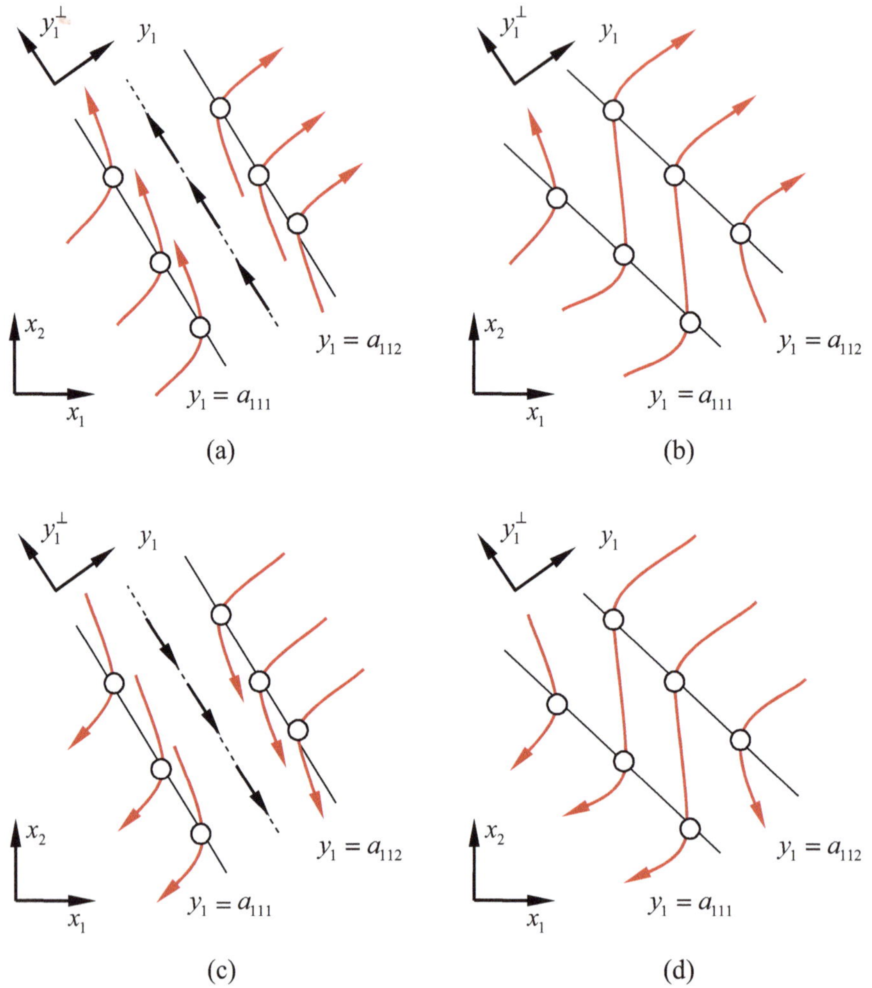

Fig. 2.6 Phase portraits for two-dimensional systems on the x_1-direction with $y_1^* = a_{111}, a_{211}$ and on the x_2-direction with $\dot{x}_2 = a_{221} \neq 0(\alpha_{11} > 0,\ \alpha_{12} > 0)$: (**a**) an (US,pF)-flow for (UP,pF) up-parabola flow and (DP,pF) down-parabola flow on the lines of $y_1^* = a_{111}, a_{112}$ $(a_{110} < 0,\ A_{222} > 0)$, (**b**) without equilibrium flow for (UP,pF) up-parabola flow and (DP,pF) down-parabola flow on the lines of $y_1^* = a_{111}, a_{112}$ $(a_{110} < 0, A_{222} > 0)$; (**c**) a (LS,nF)-flow for (DP,nF) down-parabola flow and (UP,nF) up-parabola flow on the lines of $y_1^* = a_{111}, a_{112}$ $(a_{110} < 0,\ A_{222} > 0)$, (**d**) without equilibrium flow for (DP,nF) down-parabola flow and (UP,nF) up-parabola flow on the lines of $y_1^* = a_{111}, a_{112}$ $(a_{110} < 0, A_{222} < 0)$

Chapter 3
Single-Linear-Bivariate, Linear and Quadratic Systems

In this chapter, dynamical systems with single-linear-bivariate linear and quadratic vector fields will be considered. Parabola-source and sink infinite-equilibriums for the switching of inflection and parabola flows will be discussed, and the inflection-source and sink infinite-equilibriums for parabola flows in two directions will be discussed. The inflection flows are for the appearing bifurcation of up- and down-parabola flows. The parabola flow arrays exist in such quadratic systems.

3.1 Parabola and Inflection Sources and Sinks

In this section, the dynamics of linear and quadratic single-bivariate-linear systems will be presented. The singular flows will be presented, and the corresponding switching and appearing bifurcations for the regular flows (i.e., sink, source, and parabola flows) will be discussed.

Theorem 3.1 *Consider a two-dimensional, dynamical system on the x_i-direction ($i \in \{1, 2\}$)as*

$$\begin{aligned} \dot{x}_{j_1} &= f_{j_1}(\mathbf{x}, \mathbf{p}) = A_{j_1 j_1 2} + A_{j_1 j_1 1} y_{j_1} + A_{j_1 j_1 0} y_{j_1}^2, \\ \dot{x}_{j_2} &= f_{j_2}(\mathbf{x}, \mathbf{p}) = A_{j_2 j_2 2} + A_{j_2 j_2 1} y_{j_1} + A_{j_2 j_2 0} y_{j_1}^2, \end{aligned} \tag{3.1}$$

with

$$\begin{aligned} &y_{j_1} = \alpha_{j_1 j_1} x_{j_1} + \alpha_{j_1 j_2} x_{j_2} \text{ for } j_1, j_2 \in \{1, 2\}, j_1 \neq j_2; \\ &\Delta_{jj} = (A_{jj0})^2 - 4A_{jj2}A_{jj0} \text{ for } j = j_1, j_2. \end{aligned} \tag{3.2}$$

A. C. J. Luo, *Introduction to Infinite-Equilibriums in Dynamical Systems*,
https://doi.org/10.1007/978-3-031-89083-3_3

For $A_{j_2j_2 0}=0,\ A_{j_2j_2 1}\neq 0, A_{j_2j_2 2}\neq 0,$ *the two-dimensional dynamical system is*

$$\begin{aligned}\dot{x}_{j_2} &= A_{j_2j_2 2} + A_{j_2j_2 1}y_{j_1},\\ \dot{x}_{j_1} &= A_{j_1j_1 2} + A_{j_1j_1 1}y_{j_1} + A_{j_1j_1 0}y_{j_1}^2.\end{aligned} \tag{3.3}$$

(i) If

$$A_{j_2j_2 1}\neq 0, A_{j_2j_2 2}\neq 0 \text{ and } \Delta_{j_1j_1}<0, \tag{3.4}$$

the corresponding dynamical system is

$$\begin{aligned}\dot{x}_{j_2} &= a_{j_2j_2 0}(y_{j_1} - a_{j_2j_2 1}),\\ \dot{x}_{j_1} &= a_{j_1j_1 0}\left[(y_{j_1} - a_{j_1j_1})^2 + \frac{1}{4A_{j_1j_1 0}}(-\Delta_{j_1j_1})\right],\end{aligned} \tag{3.5}$$

where

$$\begin{aligned}a_{j_2j_2 1} &= -\frac{1}{A_{j_2j_2 1}}A_{j_2j_2 2}, a_{j_2j_2 0} = A_{j_2j_2 1};\\ a_{j_1j_1} &= -\frac{1}{2A_{j_1j_1 0}}A_{j_1j_1 1}, a_{j_1j_1 0} = A_{j_1j_1 0}.\end{aligned} \tag{3.6}$$

There is a dynamical system in the y_{j_1}-direction *as*

$$\dot{y}_{j_1} = \alpha_{j_1j_1}a_{j_1j_1 0}[(y_{j_1} - a_{j_1j_1})^2 + b_{j_1j_1}] + \alpha_{j_1j_2}a_{j_2j_2 0}(y_{j_1} - a_{j_2j_2 1}), \tag{3.7}$$

with a dynamical system in the $y_{j_1}^{\perp}$*-direction as*

$$\dot{y}_{j_1}^{\perp} = \beta_{j_1j_1}a_{j_1j_1 0}[(y_{j_1} - a_{j_1j_1})^2 + b_{j_1j_1}] + \beta_{j_1j_2}a_{j_2j_2 0}(y_{j_1} - a_{j_2j_2 1}), \tag{3.8}$$

where

$$\begin{aligned}& y_{j_1}^{\perp} = \beta_{j_1j_1}x_{j_1} + \beta_{j_1j_2}x_{j_2}, \mathbf{n}_{j_1}^{\perp} = (\beta_{j_1j_1}, \beta_{j_1j_2})^{\mathrm{T}},\\ & \mathbf{n}_{j_1}\cdot\mathbf{n}_{j_1}^{\perp} = \alpha_{j_1j_1}\beta_{j_1j_1} + \alpha_{j_1j_2}\beta_{j_1j_2} = 0;\end{aligned} \tag{3.9}$$

and

$$\begin{aligned}& \Delta = (c_{j_1j_1})^2 - 4\left[b_{j_1j_1} + c_{j_1j_1}(a_{j_2j_2 1} - a_{j_1j_1 1})\right];\\ & b_{j_1j_1} = \frac{1}{4A_{j_1j_1 0}}(-\Delta_{j_1j_1}), c_{j_1j_1} = \frac{\alpha_{j_1j_1}a_{j_1j_1 0}}{\alpha_{j_1j_2}a_{j_2j_2 0}}.\end{aligned} \tag{3.10}$$

- *For $\Delta < 0$, there is no any equilibrium on the y_{j_1}- direction.*
- *For $\Delta = 0$, there is an equilibrium on the y_{j_1}- direction for the one-dimensional lower-saddle or upper-saddle flows.*
- *For $\Delta > 0$, there are two equilibriums on the y_{j_1}- direction for the one-dimensional sink and source flows.*

The simple equilibrium of $y_{j_1}^* = a_{j_2 j_2 1}$ in the x_{j_2}-direction for such a system has the following properties with:

- *For $a_{j_1 j_1 0} > 0$ and $a_{j_2 j_2 0} > 0$,*

$$(\dot{x}_{j_1}^*, a_{j_2 j_2 1}) = \underbrace{(\mathrm{pF}, \mathrm{UP})}_{\text{up-parabola flow } (+)}. \tag{3.11}$$

- *For $a_{j_1 j_1 0} < 0$ and $a_{j_2 j_2 0} > 0$,*

$$(\dot{x}_{j_1}^*, a_{j_2 j_2 1}) = \underbrace{(\mathrm{nF}, \mathrm{DP})}_{\text{down-parabola flow } (-)}. \tag{3.12}$$

- *For $a_{j_1 j_1 0} > 0$ and $a_{j_2 j_2 0} < 0$,*

$$(\dot{x}_{j_1}^*, a_{j_2 j_2 1}) = \underbrace{(\mathrm{pF}, \mathrm{DP})}_{\text{down-parabola flow } (+)}. \tag{3.13}$$

- *For $a_{j_1 j_1 0} < 0$ and $a_{j_2 j_2 0} < 0$,*

$$(\dot{x}_{j_1}^*, a_{j_2 j_2 1}) = \underbrace{(\mathrm{nF}, \mathrm{UP})}_{\text{up-parabola flow } (-)}. \tag{3.14}$$

(ii) If

$$A_{j_2 j_2 1} \neq 0, A_{j_2 j_2 2} \neq 0 \text{ and } \Delta_{j_1 j_1} = 0, \tag{3.15}$$

the two-dimensional single-bivariate dynamical system becomes

$$\begin{aligned} \dot{x}_{j_2} &= a_{j_2 j_2 0}(y_{j_1} - a_{j_2 j_2 1}), \\ \dot{x}_{j_1} &= a_{j_1 j_1 0}(y_{j_1} - a_{j_1 j_1 1})^2. \end{aligned} \tag{3.16}$$

There is a dynamical system in the y_{j_1}-direction *as*

$$\begin{aligned}\dot{y}_{j_1} = \alpha_{j_1j_1}a_{j_1j_10}[(y_{j_1}-a_{j_1j_11})^2 + c_{j_1j_1}(y_{j_1}-a_{j_1j_11})\\ +c_{j_1j_1}(a_{j_1j_11}-a_{j_2j_21})],\end{aligned} \tag{3.17}$$

with a dynamical system in the $y_{j_1}^{\perp}$-direction as

$$\dot{y}_{j_1}^{\perp} = \beta_{j_1j_1}a_{j_1j_10}(y_{j_1}-a_{j_1j_11})^2 + \beta_{j_1j_2}a_{j_2j_20}(y_{j_1}-a_{j_2j_21}), \tag{3.18}$$

where

$$\Delta = (c_{j_1j_1})^2 - 4c_{j_1j_1}(a_{j_2j_21}-a_{j_1j_11}). \tag{3.19}$$

- *For $\Delta < 0$, there is no equilibrium on the y_{j_1}-direction.*
- *For $\Delta = 0$, there is an equilibrium on the y_{j_1}-direction for the one-dimensional lower-saddle or upper-saddle flows.*
- *For $\Delta > 0$, there are two equilibriums on the y_{j_1}-direction for the one-dimensional sink and source flows.*

(ii$_1$) The equilibrium of $y_{j_1}^* = a_{j_1j_11}$ for the x_{j_1}-direction possesses the following properties with $\dot{x}_{j_2}^* = a_{j_2j_20}(a_{j_1j_11}-a_{j_2j_21})$:

- *For* $\alpha_{j_1j_2}^2 a_{j_1j_10} < 0$ and $a_{j_2j_20}(a_{j_1j_11}-a_{j_2j_21}) > 0$,

$$(a_{j_1j_11}, \dot{x}_{j_2}^*) = \underbrace{(\mathrm{II}, \mathrm{pF})}_{\text{increasing-inflection flow } (+)}. \tag{3.20}$$

- *For* $\alpha_{j_1j_2}^2 a_{j_1j_10} < 0$ and $a_{j_2j_20}(a_{j_1j_11}-a_{j_2j_21}) > 0$,

$$(a_{j_1j_11}, \dot{x}_{j_2}^*) = \underbrace{(\mathrm{DI}, \mathrm{pF})}_{\text{decreasing-inflection flow } (+)}. \tag{3.21}$$

- *For* $\alpha_{j_1j_2}^2 a_{j_1j_10} > 0$ and $a_{j_2j_20}(a_{j_1j_11}-a_{j_2j_21}) < 0$,

$$(a_{j_1j_11}, \dot{x}_{j_2}^*) = \underbrace{(\mathrm{DI}, \mathrm{nF})}_{\text{decreasing-inflection flow } (-)}. \tag{3.22}$$

- *For* $\alpha_{j_1j_2}^2 a_{j_1j_10} < 0$ and $a_{j_2j_20}(a_{j_1j_11}-a_{j_2j_21}) < 0$,

$$(a_{j_1j_11}, \dot{x}_{j_2}^*) = \underbrace{(\mathrm{II}, \mathrm{nF})}_{\text{increasing-inflection flow } (-)}. \tag{3.23}$$

(ii₂) The equilibrium of $y^*_{j_1} = a_{j_2j_21}$ *for the* x_{j_2}*-direction possesses the following properties with* $\dot{x}^*_{j_1} = a_{j_1j_10}(a_{j_2j_21} - a_{j_1j_11})^2$*:*

- For $a_{j_1j_10}(a_{j_2j_21} - a_{j_1j_11})^2 > 0$ and $\alpha_{j_1j_1}a_{j_2j_20} > 0$,

$$(\dot{x}^*_{j_1}, a_{j_2j_21}) = \underbrace{(\text{pF}, \text{UP})}_{\text{up-parabola flow } (+)}. \tag{3.24}$$

- For $a_{j_1j_10}(a_{j_2j_21} - a_{j_1j_11})^2 < 0$ and $\alpha_{j_1j_1}a_{j_2j_20} > 0$,

$$(\dot{x}^*_{j_1}, a_{j_2j_21}) = \underbrace{(\text{nF}, \text{DP})}_{\text{down-parabola flow } (-)}. \tag{3.25}$$

- For $a_{j_1j_10}(a_{j_2j_21} - a_{j_1j_11})^2 > 0$ and $\alpha_{j_1j_1}a_{j_2j_20} < 0$,

$$(\dot{x}^*_{j_1}, a_{j_2j_21}) = \underbrace{(\text{pF}, \text{DP})}_{\text{down-parabola flow } (+)}. \tag{3.26}$$

- For $a_{j_1j_10}(a_{j_2j_21} - a_{j_1j_11})^2 < 0$ and $\alpha_{j_1j_1}a_{j_2j_20} < 0$,

$$(\dot{x}^*_{j_1}, a_{j_2j_21}) = \underbrace{(\text{nF}, \text{UP})}_{\text{up-parabola flow } (-)}. \tag{3.27}$$

(ii₃) For $a_{j_2j_21} = a_{j_1j_11}$, *the two-dimensional single-bivariate dynamical system becomes*

$$\begin{aligned} \dot{x}_{j_2} &= a_{j_2j_20}(y_{j_1} - a_{j_1j_11}), \\ \dot{x}_{j_1} &= a_{j_1j_10}(y_{j_1} - a_{j_1j_11})^2. \end{aligned} \tag{3.28}$$

There is a dynamical system in the y_{j_1}-direction *as*

$$\dot{y}_{j_1} = \alpha_{j_1j_1}a_{j_1j_10}(y_{j_1} - a_{j_1j_11})(y_{j_1} - a_{j_1j_11} + c_{j_1j_1}), \tag{3.29}$$

which has two equilibriums on the y_{j_1}*- direction for the one-dimensional sink and source flows, and a dynamical system in the* $y^{\perp}_{j_1}$*-direction is*

$$\dot{y}^{\perp}_{j_1} = \beta_{j_1j_1}a_{j_1j_10}(y_{j_1} - a_{j_1j_11})^2 + \beta_{j_1j_2}a_{j_2j_20}(y_{j_1} - a_{j_1j_11}). \tag{3.30}$$

The infinite-equilibrium of $y^*_{j_1} = a_{j_1j_11}$ *for the* x_{j_1}*-direction and* $y^*_{j_1} = a_{j_2j_21}$ *for the* x_{j_2}*-direction possess the following properties.*

- For $\alpha^2_{j_1j_2}a_{j_1j_1 0} > 0$ and $\alpha_{j_1j_2}a_{j_2j_2 0} > 0$,

$$(a_{j_1j_1 1}, a_{j_2j_2 1}) = \underbrace{(\mathrm{UP}, \mathrm{SO})}_{\text{infinite-equilibrium}}. \tag{3.31}$$

- For $\alpha^2_{j_1j_2}a_{j_1j_1 0} < 0$ and $\alpha_{j_1j_2}a_{j_2j_2 0} > 0$,

$$(a_{j_1j_1 1}, a_{j_2j_2 1}) = \underbrace{(\mathrm{DP}, \mathrm{SO})}_{\text{infinite-equilibrium}}. \tag{3.32}$$

- For $\alpha^2_{j_1j_2}a_{j_1j_1 0} > 0$ and $\alpha_{j_1j_2}a_{j_2j_2 0} < 0$,

$$(a_{j_1j_1 1}, a_{j_2j_2 1}) = \underbrace{(\mathrm{DP}, \mathrm{SI})}_{\text{infinite-equilibrium}}. \tag{3.33}$$

- For $\alpha^2_{j_1j_2}a_{j_1j_1 0} < 0$ and $\alpha_{j_1j_2}a_{j_2j_2 0} < 0$,

$$(a_{j_1j_1 1}, a_{j_2j_2 1}) = \underbrace{(\mathrm{UP}, \mathrm{SI})}_{\text{infinite-equilibrium}}. \tag{3.34}$$

(ii$_4$) For $a_{j_2j_2 1} = a_{j_1j_1 1}$, the infinite-equilibrium of $y^*_{j_1} = a_{j_1j_1 1}$ is a switching bifurcation.

- *For* $a_{j_1j_1 0} > 0$ and $a_{j_2j_2 0} > 0$,

$$\left\{\begin{array}{c}\underbrace{(\mathrm{pF}, \mathrm{UP})}_{\text{up-parabola flow }(+)}\\ \underbrace{(\mathrm{II}, \mathrm{pF})}_{\text{increasing-inflection flow }(+)}\end{array}\right\} \rightleftharpoons \underbrace{(\mathrm{UP}, \mathrm{SO})}_{\text{switching bifurcation}} \rightleftharpoons \left\{\begin{array}{c}\underbrace{(\mathrm{DI}, \mathrm{nF})}_{\text{decreasing-inflection flow }(-)}\\ \underbrace{(\mathrm{pF}, \mathrm{UP})}_{\text{up-parabola flow }(+)}\end{array}\right\}$$

for $\alpha_{j_1j_2} > 0$;

$$\left\{\begin{array}{c}\underbrace{(\mathrm{pF}, \mathrm{UP})}_{\text{up-parabola flow }(+)}\\ \underbrace{(\mathrm{II}, \mathrm{pF})}_{\text{increasing-inflection flow }(+)}\end{array}\right\} \rightleftharpoons \underbrace{(\mathrm{DP}, \mathrm{SI})}_{\text{switching-bifurcation}} \rightleftharpoons \left\{\begin{array}{c}\underbrace{(\mathrm{DI}, \mathrm{nF})}_{\text{decreasing-inflection flow }(-)}\\ \underbrace{(\mathrm{pF}, \mathrm{UP})}_{\text{up-parabola flow }(+)}\end{array}\right\}$$

for $\alpha_{j_1j_2} < 0$.

(3.35)

- *For* $a_{j_1j_10} > 0$ and $a_{j_2j_20} < 0$,

$$\left\{\begin{matrix}\underbrace{(\mathrm{nF},\mathrm{DP})}_{\text{down-parabola flow }(-)}\\ \underbrace{(\mathrm{DI},\mathrm{pF})}_{\text{decreasing-inflection flow }(+)}\end{matrix}\right\} \rightleftharpoons \underbrace{(\mathrm{DP},\mathrm{SO})}_{\text{switching-bifurcation}} \rightleftharpoons \left\{\begin{matrix}\underbrace{(\mathrm{II},\mathrm{nF})}_{\text{increasing-inflection flow }(-)}\\ \underbrace{(\mathrm{nF},\mathrm{DP})}_{\text{down-parabola flow }(-)}\end{matrix}\right\}$$
for $\alpha_{j_1j_2} > 0$;

$$\left\{\begin{matrix}\underbrace{(\mathrm{nF},\mathrm{DP})}_{\text{down-parabola flow }(-)}\\ \underbrace{(\mathrm{DI},\mathrm{pF})}_{\text{decreasing-inflection flow }(+)}\end{matrix}\right\} \rightleftharpoons \underbrace{(\mathrm{UP},\mathrm{SI})}_{\text{switching-bifurcation}} \rightleftharpoons \left\{\begin{matrix}\underbrace{(\mathrm{II},\mathrm{nF})}_{\text{increasing-inflection flow }(-)}\\ \underbrace{(\mathrm{nF},\mathrm{DP})}_{\text{down-parabola flow }(-)}\end{matrix}\right\}$$
for $\alpha_{j_1j_2} > 0$.

(3.36)

- *For* $a_{j_1j_10} < 0$ and $a_{j_2j_20} > 0$,

$$\left\{\begin{matrix}\underbrace{(\mathrm{pF},\mathrm{DP})}_{\text{down-parabola flow }(+)}\\ \underbrace{(\mathrm{DI},\mathrm{nF})}_{\text{decreasing-inflection flow }(-)}\end{matrix}\right\} \rightleftharpoons \underbrace{(\mathrm{DP},\mathrm{SI})}_{\text{switching-bifurcation}} \rightleftharpoons \left\{\begin{matrix}\underbrace{(\mathrm{II},\mathrm{pF})}_{\text{increasing-inflection flow }(+)}\\ \underbrace{(\mathrm{pF},\mathrm{DP})}_{\text{down-parabola flow }(+)}\end{matrix}\right\}$$
for $\alpha_{j_1j_2} > 0$;

$$\left\{\begin{matrix}\underbrace{(\mathrm{pF},\mathrm{DP})}_{\text{down-parabola flow }(+)}\\ \underbrace{(\mathrm{DI},\mathrm{nF})}_{\text{decreasing-inflection flow }(-)}\end{matrix}\right\} \rightleftharpoons \underbrace{(\mathrm{UP},\mathrm{SO})}_{\text{switching-bifucations}} \rightleftharpoons \left\{\begin{matrix}\underbrace{(\mathrm{II},\mathrm{pF})}_{\text{increasing-inflection flow }(+)}\\ \underbrace{(\mathrm{pF},\mathrm{DP})}_{\text{down-parabola flow }(+)}\end{matrix}\right\}$$
for $\alpha_{j_1j_2} < 0$.

(3.37)

- *For* $a_{j_1j_10}<0$ and $a_{j_2j_20}<0$,

$$\left\{\begin{array}{c}\underbrace{(\mathrm{nF},\mathrm{UP})}_{\text{up-parabola flow }(-)}\\ \underbrace{(\mathrm{II},\mathrm{nF})}_{\text{increasing-inflection flow }(-)}\end{array}\right\}\rightleftharpoons\underbrace{(\mathrm{UP},\mathrm{SI})}_{\text{switching-bifurcation}}\rightleftharpoons\left\{\begin{array}{c}\underbrace{(\mathrm{DI},\mathrm{pF})}_{\text{decreasing-inflection flow }(+)}\\ \underbrace{(\mathrm{nF},\mathrm{UP})}_{\text{up-parabola flow }(-)}\end{array}\right\}$$

for $\alpha_{j_1j_2}>0$;

$$\left\{\begin{array}{c}\underbrace{(\mathrm{nF},\mathrm{UP})}_{\text{up-parabola flow }(-)}\\ \underbrace{(\mathrm{II},\mathrm{nF})}_{\text{increasing-inflection flow }(-)}\end{array}\right\}\rightleftharpoons\underbrace{(\mathrm{DP},\mathrm{SO})}_{\text{switching-bifurcation}}\rightleftharpoons\left\{\begin{array}{c}\underbrace{(\mathrm{DI},\mathrm{pF})}_{\text{decreasing-inflection flow }(+)}\\ \underbrace{(\mathrm{nF},\mathrm{UP})}_{\text{up-parabola flow }(-)}\end{array}\right\}$$

for $\alpha_{j_1j_2}<0$.

(3.38)

(iii) If

$$A_{j_2j_21}\neq 0, A_{j_2j_22}\neq 0 \text{ and } \Delta_{j_1j_1}>0, \tag{3.39}$$

the two-dimensional dynamical system has two simple equilibriums for the y_{j_1}*-direction on the* x_i*-direction, i.e.,*

$$\begin{aligned} y_{j_1}^* &= a_{j_1j_11}, a_{j_1j_12} \text{ for the } x_{j_1}\text{-direction},\\ y_{j_1}^* &= a_{j_2j_21} \text{ for the } x_{j_2}\text{-direction}, \end{aligned} \tag{3.40}$$

where

$$\begin{aligned} &\{a_{j_1j_11}, a_{j_1j_12}\} = \mathrm{sort}\{b_{j_1j_11}, b_{j_1j_12}\}, a_{j_1j_11}<a_{j_1j_12};\\ &b_{j_1j_11}, b_{j_1j_12} = -\frac{1}{2A_{j_1j_10}}\left(A_{j_1j_11}\pm\sqrt{\Delta_{j_1j_1}}\right), a_{j_1j_10}=A_{j_1j_10}, \end{aligned} \tag{3.41}$$

and the corresponding dynamical system becomes

$$\begin{aligned} \dot{x}_{j_2} &= a_{j_2j_20}(y_{j_1}-a_{j_2j_21}),\\ \dot{x}_{j_1} &= a_{j_1j_10}(y_{j_1}-a_{j_1j_11})(y_{j_1}-a_{j_1j_12}). \end{aligned} \tag{3.42}$$

There is a dynamical system in the y_{j_1}-direction *as*

$$\begin{aligned}\dot{y}_{j_1} &= \alpha_{j_1j_1}a_{j_1j_10}[(y_{j_1}-a_{j_1j_11})(y_{j_1}-a_{j_1j_12})+c_{j_1j_1}(y_{j_1}-a_{j_2j_21})] \\ &= \alpha_{j_1j_1}a_{j_1j_10}[y_{j_1}^2-(a_{j_1j_11}+a_{j_1j_12}-c_{j_1j_1})y_{j_1} \\ &\quad +(a_{j_1j_11}a_{j_1j_12}-c_{j_1j_1}a_{j_2j_21})],\end{aligned} \tag{3.43}$$

with a dynamical system in the $y_{j_1}^{\perp}$-direction as

$$\begin{aligned}\dot{y}_{j_1}^{\perp} = &\ \beta_{j_1j_1}a_{j_1j_10}(y_{j_1}-a_{j_1j_11})(y_{j_1}-a_{j_1j_12}) \\ &+\beta_{j_1j_2}a_{j_2j_20}(y_{j_1}-a_{j_2j_21}),\end{aligned} \tag{3.44}$$

and

$$\Delta=(a_{j_1j_11}+a_{j_1j_12}-c_{j_1j_1})^2-4(a_{j_1j_11}a_{j_1j_12}-c_{j_1j_1}a_{j_2j_21}). \tag{3.45}$$

- *For $\Delta<0$, there is no any equilibrium on the y_{j_1}-direction.*
- *For $\Delta=0$, there is an equilibrium on the y_{j_1}-direction for the one-dimensional lower-saddle or upper-saddle flows.*
- *For $\Delta>0$, there are two equilibriums on the y_{j_1}-direction for the one-dimensional sink and source flows.*

(iii$_1$) The equilibrium of $y_{j_1}^*=a_{j_1j_1s}$ (s, $l\in\{1,2\}, s\neq l$) for the x_{j_1}-direction with $\dot{x}_{j_2}^*=a_{j_2j_20}(a_{j_1j_1s}-a_{j_2j_21})$ possesses the following properties:

- For $a_{j_1j_10}\alpha_{j_1j_2}(a_{j_1j_1s}-a_{j_1j_1l})>0$ and $a_{j_2j_20}(a_{j_1j_1s}-a_{j_2j_21})>0$,

$$(a_{j_1j_1s},\dot{x}_{j_2}^*)=\underbrace{(\mathrm{UP},\mathrm{pF})}_{\text{up-parabola flow }(+)}. \tag{3.46}$$

- For $a_{j_1j_10}\alpha_{j_1j_2}(a_{j_1j_1s}-a_{j_1j_1l})<0$ and $a_{j_2j_20}(a_{j_1j_1s}-a_{j_2j_21})>0$,

$$(a_{j_1j_1s},\dot{x}_{j_2}^*)=\underbrace{(\mathrm{DP},\mathrm{pF})}_{\text{down-parabola flow }(+)}. \tag{3.47}$$

- For $a_{j_1j_10}\alpha_{j_1j_2}(a_{j_1j_1s}-a_{j_1j_1l})>0$ and $a_{j_2j_20}(a_{j_1j_1s}-a_{j_2j_21})<0$,

$$(a_{j_1j_1s},\dot{x}_{j_2}^*)=\underbrace{(\mathrm{DP},\mathrm{nF})}_{\text{down-parabola flow }(-)}. \tag{3.48}$$

- For $a_{j_1j_10}\alpha_{j_1j_2}(a_{j_1j_1s}-a_{j_1j_1l})<0$ and $a_{j_2j_20}(a_{j_1j_1s}-a_{j_2j_21})<0$,

$$(a_{j_1j_1s},\dot{x}_{j_2}^*)=\underbrace{(\mathrm{UP},\mathrm{nF})}_{\text{up-parabola flow }(-)}. \tag{3.49}$$

(iii$_2$) The equilibrium of $y_{j_1}^* = a_{j_2j_21}$ for the x_{j_2}-direction possesses the following properties with $\dot{x}_{j_1}^* = a_{j_1j_10}\prod_{s=1}^{2}(a_{j_2j_21} - a_{j_1j_1s})$:

- For $a_{j_1j_10}\prod_{s=1}^{2}(a_{j_2j_21} - a_{j_1j_1s}) > 0$ and $a_{j_2j_20}\alpha_{j_1j_1} > 0$,

$$(\dot{x}_{j_1}^*, a_{j_2j_21}) = \underbrace{(\text{pF}, \text{UP})}_{\text{up-parabola flow } (+)}. \tag{3.50}$$

- For $a_{j_1j_10}\prod_{s=1}^{2}(a_{j_2j_21} - a_{j_1j_1s}) < 0$ and $a_{j_2j_20}\alpha_{j_1j_1} > 0$,

$$(\dot{x}_{j_1}^*, a_{j_2j_21}) = \underbrace{(\text{nF}, \text{DP})}_{\text{down-parabola flow } (-)}. \tag{3.51}$$

- For $a_{j_1j_10}\prod_{s=1}^{2}(a_{j_2j_21} - a_{j_1j_1s}) > 0$ and $a_{j_2j_20}\alpha_{j_1j_1} < 0$,

$$(\dot{x}_{j_1}^*, a_{j_2j_21}) = \underbrace{(\text{pF}, \text{DP})}_{\text{down-parabola flow } (+)}. \tag{3.52}$$

- For $a_{j_1j_10}\prod_{s=1}^{2}(a_{j_2j_21} - a_{j_1j_1s}) < 0$ and $a_{j_2j_20}\alpha_{j_1j_1} < 0$,

$$(\dot{x}_{j_1}^*, a_{j_2j_21}) = \underbrace{(\text{nF}, \text{UP})}_{\text{up-parabola flow } (-)}. \tag{3.53}$$

(iv) For $a_{j_2j_21} = a_{j_1j_1s}(s, l \in \{1,2\}, s \neq l)$, the corresponding dynamical system becomes

$$\begin{aligned} \dot{x}_{j_2} &= a_{j_2j_20}(y_{j_1} - a_{j_2j_21}), \\ \dot{x}_{j_1} &= a_{j_1j_10}(y_{j_1} - a_{j_1j_1s})(y_{j_1} - a_{j_1j_1l}). \end{aligned} \tag{3.54}$$

There is a dynamical system in the y_{j_1}-direction *as*

$$\dot{y}_{j_1} = \alpha_{j_1j_1}a_{j_1j_10}(y_{j_1} - a_{j_1j_1s})(y_{j_1} - a_{j_1j_1l} + c_{j_1j_1}), \tag{3.55}$$

with a dynamical system in the $y_{j_1}^{\perp}$-direction as

$$\dot{y}_{j_1}^{\perp} = \beta_{j_1j_1}a_{j_1j_10}(y_{j_1} - a_{j_1j_1s})(y_{j_1} - a_{j_1j_12} + c_{j_1j_11}), \tag{3.56}$$

where

$$c_{j_1j_11}=\frac{\beta_{j_1j_2}a_{j_2j_20}}{\beta_{j_1j_1}a_{j_1j_10}}, G_{j_1j_10}=\alpha_{j_1j_1}a_{j_1j_10}(a_{j_1j_1s}-a_{j_1j_1l})+\alpha_{j_1j_2}a_{j_2j_20}. \tag{3.57}$$

The equilibrium of $y_{j_1}=a_{j_1j_1s}$ *is an infinite-equilibrium.*
(iv$_1$) For $a_{j_1j_1s}=a_{j_2j_21}$*, the infinite-equilibrium of* $y^*_{j_1}=a_{j_1j_1s}$ $(s, l\in\{1,2\}, s\neq l)$ *for the* x_{j_1} *-direction and* $y^*_{j_1}=a_{j_2j_21}$ *for the* x_{j_2} *-direction possesses the following properties:*

- For $a_{j_1j_10}(a_{j_1j_1s}-a_{j_1j_1l})/a_{j_2j_20}>0$ and $G_{j_1j_10}>0$,

$$(a_{j_1j_1s},a_{j_2j_21})=\underbrace{(\mathrm{II},\mathrm{SO})}_{\text{increasing-inflection source}}. \tag{3.58}$$

- For $a_{j_1j_10}(a_{j_1j_1s}-a_{j_1j_1l})/a_{j_2j_20}>0$ and $G_{j_1j_10}<0$,

$$(a_{j_1j_1s},a_{j_2j_21})=\underbrace{(\mathrm{II},\mathrm{SI})}_{\text{increasing-inflection sink}}. \tag{3.59}$$

- For $a_{j_1j_10}(a_{j_1j_1s}-a_{j_1j_1l})/a_{j_2j_20}<0$ and $G_{j_1j_10}>0$,

$$(a_{j_1j_1s},a_{j_2j_21})=\underbrace{(\mathrm{DI},\mathrm{SO})}_{\text{decreasing-inflection source}}. \tag{3.60}$$

- For $a_{j_1j_10}(a_{j_1j_1s}-a_{j_1j_1l})/a_{j_2j_20}<0$ and $G_{j_1j_10}<0$,

$$(a_{j_1j_1s},a_{j_2j_21})=\underbrace{(\mathrm{DI},\mathrm{SI})}_{\text{decreasing-inflection sink}}. \tag{3.61}$$

(iv$_2$) The equilibrium of $y^*_{j_1}=a_{j_1j_1l}$ $(s, l\in\{1,2\}, l\neq s)$ *for the* x_{j_1} *-direction with* $\dot{x}^*_{j_2}=a_{j_2j_20}(a_{j_1j_1l}-a_{j_2j_21})$ *possesses the following properties:*

- For $\alpha_{j_1j_2}a_{j_1j_10}(a_{j_1j_1l}-a_{j_1j_1s})>0$ and $a_{j_2j_20}(a_{j_1j_1l}-a_{j_2j_21})>0$,

$$(a_{j_1j_1l},\dot{x}^*_{j_2})=\underbrace{(\mathrm{UP},\mathrm{pF})}_{\text{up-parabola flow }(+)}. \tag{3.62}$$

- For $\alpha_{j_1j_2}a_{j_1j_10}(a_{j_1j_1l}-a_{j_1j_1s})<0$ and $a_{j_2j_20}(a_{j_1j_1l}-a_{j_2j_21})>0$,

$$(a_{j_1j_1l},\dot{x}^*_{j_2})=\underbrace{(\mathrm{DP},\mathrm{pF})}_{\text{down-parabola flow }(+)}. \tag{3.63}$$

- For $\alpha_{j_1j_2}a_{j_1j_10}(a_{j_1j_1l}-a_{j_1j_1s})>0$ and $a_{j_2j_20}(a_{j_1j_1l}-a_{j_2j_21})<0$,

$$(a_{j_1j_1l},\dot{x}_{j_2}^{*})=\underbrace{(\mathrm{DP},\mathrm{nF})}_{\text{down-parabola flow }(-)}. \tag{3.64}$$

- For $\alpha_{j_1j_2}a_{j_1j_10}(a_{j_1j_1l}-a_{j_1j_1s})<0$ and $a_{j_2j_20}(a_{j_1j_1l}-a_{j_2j_21})<0$,

$$(a_{j_1j_1l},\dot{x}_{j_2}^{*})=\underbrace{(\mathrm{UP},\mathrm{nF})}_{\text{up-parabola flow }(-)}. \tag{3.65}$$

(iv₃) For $y_{j_1}^{*}=a_{j_1j_1s}=a_{j_2j_21}$ $(s,\ l\in\{1,2\},\ l\neq s)$, the infinite-equilibrium of $(a_{j_1j_1s},a_{j_2j_21})$ is a switching bifurcation for $\alpha_{j_1j_1}>0$.

- For $a_{j_1j_10}(a_{j_1j_1s}-a_{j_1j_1l})>0$ and $a_{j_2j_20}>0$,

$$\left\{\begin{array}{c}\underbrace{(\mathrm{nF},\mathrm{DP})}_{\text{down-parabola flow }(-)}\\ \underbrace{(\mathrm{UP},\mathrm{pF})}_{\text{up-parabola flow }(+)}\end{array}\right\}\rightleftharpoons\underbrace{(\mathrm{II},\mathrm{SO})}_{\text{switching bifurcation}}\rightleftharpoons\left\{\begin{array}{c}\underbrace{(\mathrm{DP},\mathrm{nF})}_{\text{down-parabola flow }(-)}\\ \underbrace{(\mathrm{pF},\mathrm{UP})}_{\text{up-parabola flow }(+)}\end{array}\right\}$$
for $\alpha_{j_1j_2}>0, G_{j_1j_10}>0$;

$$\left\{\begin{array}{c}\underbrace{(\mathrm{nF},\mathrm{DP})}_{\text{down-parabola flow }(-)}\\ \underbrace{(\mathrm{DP},\mathrm{pF})}_{\text{down-parabola flow }(+)}\end{array}\right\}\rightleftharpoons\underbrace{(\mathrm{II},\mathrm{SO})}_{\text{switching bifurcation}}\rightleftharpoons\left\{\begin{array}{c}\underbrace{(\mathrm{UP},\mathrm{nF})}_{\text{up-parabola flow }(-)}\\ \underbrace{(\mathrm{pF},\mathrm{UP})}_{\text{up-parabola flow }(+)}\end{array}\right\} \tag{3.66}$$
for $\alpha_{j_1j_2}<0, G_{j_1j_10}>0$;

$$\left\{\begin{array}{c}\underbrace{(\mathrm{nF},\mathrm{DP})}_{\text{down-parabola flow }(-)}\\ \underbrace{(\mathrm{DP},\mathrm{pF})}_{\text{down-parabola flow }(+)}\end{array}\right\}\rightleftharpoons\underbrace{(\mathrm{II},\mathrm{SI})}_{\text{switching bifurcation}}\rightleftharpoons\left\{\begin{array}{c}\underbrace{(\mathrm{UP},\mathrm{nF})}_{\text{up-parabola flow }(-)}\\ \underbrace{(\mathrm{pF},\mathrm{UP})}_{\text{up-parabola flow }(+)}\end{array}\right\}$$
for $\alpha_{j_1j_2}<0, G_{j_1j_10}<0$.

- For $a_{j_1 j_1 0}(a_{j_1 j_1 s} - a_{j_1 j_1 l}) < 0$ and $a_{j_2 j_2 0} > 0$,

$$
\begin{aligned}
&\left\{\begin{array}{c}\underbrace{(\mathrm{pF},\mathrm{UP})}_{\text{up-parabola flow }(+)}\\ \underbrace{(\mathrm{DP},\mathrm{pF})}_{\text{down-parabola flow }(+)}\end{array}\right\} \rightleftharpoons \underbrace{(\mathrm{DI},\mathrm{SI})}_{\text{switching bifurcation}} \rightleftharpoons \left\{\begin{array}{c}\underbrace{(\mathrm{UP},\mathrm{nF})}_{\text{up-parabola flow }(-)}\\ \underbrace{(\mathrm{nF},\mathrm{DP})}_{\text{down-parabola flow }(-)}\end{array}\right\}\\
&\text{for } \alpha_{j_1 j_2} > 0, G_{j_1 j_1 0} < 0;\\
&\left\{\begin{array}{c}\underbrace{(\mathrm{pF},\mathrm{UP})}_{\text{up-parabola flow }(+)}\\ \underbrace{(\mathrm{DP},\mathrm{pF})}_{\text{down-parabola flow }(+)}\end{array}\right\} \rightleftharpoons \underbrace{(\mathrm{DI},\mathrm{SO})}_{\text{switching bifurcation}} \rightleftharpoons \left\{\begin{array}{c}\underbrace{(\mathrm{UP},\mathrm{nF})}_{\text{up-parabola flow }(-)}\\ \underbrace{(\mathrm{nF},\mathrm{DP})}_{\text{down-parabola flow }(-)}\end{array}\right\}\\
&\text{for } \alpha_{j_1 j_2} > 0, G_{j_1 j_1 0} > 0;\\
&\left\{\begin{array}{c}\underbrace{(\mathrm{pF},\mathrm{UP})}_{\text{up-parabola flow }(+)}\\ \underbrace{(\mathrm{UP},\mathrm{pF})}_{\text{up-parabola flow }(+)}\end{array}\right\} \rightleftharpoons \underbrace{(\mathrm{DI},\mathrm{SI})}_{\text{siwthcing bifurcation}} \rightleftharpoons \left\{\begin{array}{c}\underbrace{(\mathrm{DP},\mathrm{nF})}_{\text{down-parabola flow }(-)}\\ \underbrace{(\mathrm{nF},\mathrm{DP})}_{\text{down-parabola flow }(-)}\end{array}\right\}\\
&\text{for } \alpha_{j_1 j_2} < 0, G_{j_1 j_1 0} < 0.
\end{aligned}
\tag{3.67}
$$

- For $a_{j_1 j_1 0}(a_{j_1 j_1 s} - a_{j_1 j_1 l}) > 0$ and $a_{j_2 j_2 0} < 0$,

$$
\begin{aligned}
&\left\{\begin{array}{c}\underbrace{(\mathrm{nF},\mathrm{UP})}_{\text{up-parabola flow }(-)}\\ \underbrace{(\mathrm{DP},\mathrm{nF})}_{\text{down-parabola flow }(-)}\end{array}\right\} \rightleftharpoons \underbrace{(\mathrm{DI},\mathrm{SO})}_{\text{switching bifurcartion}} \rightleftharpoons \left\{\begin{array}{c}\underbrace{(\mathrm{UP},\mathrm{pF})}_{\text{up-parabola flow }(+)}\\ \underbrace{(\mathrm{pF},\mathrm{DP})}_{\text{down-parabola flow }(+)}\end{array}\right\}\\
&\text{for } \alpha_{j_1 j_2} > 0, G_{j_1 j_1 0} > 0;\\
&\left\{\begin{array}{c}\underbrace{(\mathrm{nF},\mathrm{UP})}_{\text{up-parabola flow }(-)}\\ \underbrace{(\mathrm{DP},\mathrm{nF})}_{\text{down-parabola flow }(-)}\end{array}\right\} \rightleftharpoons \underbrace{(\mathrm{DI},\mathrm{SI})}_{\text{switching bifurcartion}} \rightleftharpoons \left\{\begin{array}{c}\underbrace{(\mathrm{UP},\mathrm{pF})}_{\text{up-parabola flow }(+)}\\ \underbrace{(\mathrm{pF},\mathrm{DP})}_{\text{down-parabola flow }(+)}\end{array}\right\}\\
&\text{for } \alpha_{j_1 j_2} > 0, G_{j_1 j_1 0} < 0;\\
&\left\{\begin{array}{c}\underbrace{(\mathrm{nF},\mathrm{UP})}_{\text{up-parabola flow }(-)}\\ \underbrace{(\mathrm{UP},\mathrm{nF})}_{\text{up-parabola flow }(-)}\end{array}\right\} \rightleftharpoons \underbrace{(\mathrm{DI},\mathrm{SO})}_{\text{switching bifurcation}} \rightleftharpoons \left\{\begin{array}{c}\underbrace{(\mathrm{DP},\mathrm{pF})}_{\text{down-parabola flow }(+)}\\ \underbrace{(\mathrm{pF},\mathrm{DP})}_{\text{down-parabola flow }(+)}\end{array}\right\}\\
&\text{for } \alpha_{j_1 j_2} < 0, G_{j_1 j_1 0} > 0.
\end{aligned}
\tag{3.68}
$$

- For $a_{j_1j_10}(a_{j_1j_1s}-a_{j_1j_1l})<0$ and $a_{j_2j_20}<0$,

$$\left\{\begin{array}{c}\underbrace{(\mathrm{pF},\mathrm{DP})}_{\text{down-parabola flow }(+)}\\ \underbrace{(\mathrm{UP},\mathrm{nF})}_{\text{up-parabola flow }(-)}\end{array}\right\}\rightleftharpoons\underbrace{(\mathrm{II},\mathrm{SO})}_{\text{switching bifurcation}}\rightleftharpoons\left\{\begin{array}{c}\underbrace{(\mathrm{DP},\mathrm{pF})}_{\text{down-parabola flow }(+)}\\ \underbrace{(\mathrm{nF},\mathrm{UP})}_{\text{up-parabola flow }(-)}\end{array}\right\}$$
$$\text{for } \alpha_{j_1j_2}>0, G_{j_1j_10}<0;$$
$$\left\{\begin{array}{c}\underbrace{(\mathrm{pF},\mathrm{DP})}_{\text{down-parabola flow }(+)}\\ \underbrace{(\mathrm{DP},\mathrm{nF})}_{\text{down-parabola flow }(-)}\end{array}\right\}\rightleftharpoons\underbrace{(\mathrm{II},\mathrm{SI})}_{\text{switching bifurcation}}\rightleftharpoons\left\{\begin{array}{c}\underbrace{(\mathrm{UP},\mathrm{pF})}_{\text{up-parabola flow }(+)}\\ \underbrace{(\mathrm{nF},\mathrm{UP})}_{\text{up-parabola flow }(-)}\end{array}\right\} \quad (3.69)$$
$$\text{for } \alpha_{j_1j_2}<0, G_{j_1j_10}<0;$$
$$\left\{\begin{array}{c}\underbrace{(\mathrm{pF},\mathrm{DP})}_{\text{down-parabola flow }(+)}\\ \underbrace{(\mathrm{DP},\mathrm{nF})}_{\text{down-parabola flow }(-)}\end{array}\right\}\rightleftharpoons\underbrace{(\mathrm{II},\mathrm{SO})}_{\text{switching bifurcation}}\rightleftharpoons\left\{\begin{array}{c}\underbrace{(\mathrm{UP},\mathrm{pF})}_{\text{up-parabola flow }(+)}\\ \underbrace{(\mathrm{nF},\mathrm{UP})}_{\text{up-parabola flow }(-)}\end{array}\right\}$$
$$\text{for } \alpha_{j_1j_2}<0, G_{j_1j_10}>0.$$

(v) Under

$$A_{j_2j_21}\neq 0, A_{j_2j_22}\neq 0 \text{ and } \Delta_{j_1j_1}=0, \quad (3.70)$$

there is a bifurcation state with

$$\begin{aligned}\dot{x}_{j_2}&=a_{j_2j_20}(y_{j_1}-a_{j_2j_21}),\\ \dot{x}_{j_1}&=a_{j_1j_10}(y_{j_1}-a_{j_1j_11})^2.\end{aligned} \quad (3.71)$$

The equilibrium of $y^*_{j_1}=a_{j_1j_11}$ *for the* x_{j_1}*-direction has four switching bifurcations.*

- For $a_{j_1 j_1 0} > 0$ and $a_{j_2 j_2 0} > 0$,

$$\underbrace{(\mathrm{pF}, \mathrm{UP})}_{\text{up-parabola flow } (+)} \rightleftharpoons \underbrace{(\mathrm{UP}, \mathrm{SO})}_{\text{switching bifurcation}} \rightleftharpoons \left\{ \begin{array}{c} \underbrace{(\mathrm{UP}, \mathrm{pF})}_{\text{up-parabola flow } (+)} \\ \underbrace{(\mathrm{nF}, \mathrm{DP})}_{\text{down-parabola flow } (-)} \\ \underbrace{(\mathrm{UP}, \mathrm{nF})}_{\text{up-parabola flow } (-)} \end{array} \right\} \text{for } \alpha_{j_1 j_2} > 0,$$

$$\underbrace{(\mathrm{pF}, \mathrm{UP})}_{\text{up-parabola flow } (+)} \rightleftharpoons \underbrace{(\mathrm{DP}, \mathrm{SI})}_{\text{switching bifurcation}} \rightleftharpoons \left\{ \begin{array}{c} \underbrace{(\mathrm{DP}, \mathrm{pF})}_{\text{down-parabola flow } (+)} \\ \underbrace{(\mathrm{nF}, \mathrm{DP})}_{\text{down-parabola flow } (-)} \\ \underbrace{(\mathrm{DP}, \mathrm{nF})}_{\text{down-parabola flow } (-)} \end{array} \right\} \text{for } \alpha_{j_1 j_2} < 0. \tag{3.72}$$

- For $a_{j_1 j_1 0} < 0$ and $a_{j_2 j_2 0} > 0$,

$$\underbrace{(\mathrm{nF}, \mathrm{DP})}_{\text{down-parabola flow } (-)} \rightleftharpoons \underbrace{(\mathrm{DP}, \mathrm{SO})}_{\text{switching bifurcation}} \rightleftharpoons \left\{ \begin{array}{c} \underbrace{(\mathrm{DP}, \mathrm{nF})}_{\text{down-parabola flow } (-)} \\ \underbrace{(\mathrm{pF}, \mathrm{UP})}_{\text{up-parabola flow } (+)} \\ \underbrace{(\mathrm{DP}, \mathrm{nF})}_{\text{down-parabola flow } (-)} \end{array} \right\} \text{for } \alpha_{j_1 j_2} > 0,$$

$$\underbrace{(\mathrm{nF}, \mathrm{DP})}_{\text{down-parabola flow } (-)} \rightleftharpoons \underbrace{(\mathrm{UP}, \mathrm{SI})}_{\text{switching bifurcation}} \rightleftharpoons \left\{ \begin{array}{c} \underbrace{(\mathrm{UP}, \mathrm{pF})}_{\text{up-parabola flow } (+)} \\ \underbrace{(\mathrm{pF}, \mathrm{UP})}_{\text{up-parabola flow } (+)} \\ \underbrace{(\mathrm{UP}, \mathrm{nF})}_{\text{up-parabola flow } (-)} \end{array} \right\} \text{for } \alpha_{j_1 j_2} < 0. \tag{3.73}$$

- For $a_{j_1j_10}>0$ and $a_{j_2j_20}<0$,

$$\underbrace{(\text{pF},\text{DP})}_{\text{down-parabola flow }(-)} \rightleftharpoons \underbrace{(\text{DP},\text{SI})}_{\text{switching bifurcation}} \rightleftharpoons \left\{\begin{array}{c}\underbrace{(\text{DP},\text{nF})}_{\text{down-parabola flow }(-)}\\ \underbrace{(\text{nF},\text{UP})}_{\text{up-parabola flow }(-)}\\ \underbrace{(\text{DP},\text{pF})}_{\text{down-parabola flow }(+)}\end{array}\right\} \text{for } \alpha_{j_1j_2}>0,$$

$$\underbrace{(\text{pF},\text{DP})}_{\text{down-parabola flow }(-)} \rightleftharpoons \underbrace{(\text{UP},\text{SO})}_{\text{switching bifurcation}} \rightleftharpoons \left\{\begin{array}{c}\underbrace{(\text{UP},\text{nF})}_{\text{up-parabola flow }(-)}\\ \underbrace{(\text{nF},\text{UP})}_{\text{up-parabola flow }(-)}\\ \underbrace{(\text{UP},\text{pF})}_{\text{up-parabola flow }(+)}\end{array}\right\} \text{for } \alpha_{j_1j_2}<0. \tag{3.74}$$

- For $a_{j_1j_10}<0$ and $a_{j_2j_20}<0$,

$$\underbrace{(\text{nF},\text{UP})}_{\text{up-parabola flow }(-)} \rightleftharpoons \underbrace{(\text{UP},\text{SI})}_{\text{switching bifurcation}} \rightleftharpoons \left\{\begin{array}{c}\underbrace{(\text{UP},\text{nF})}_{\text{up-parabola flow }(-)}\\ \underbrace{(\text{pF},\text{DP})}_{\text{down-parabola flow }(+)}\\ \underbrace{(\text{UP},\text{pF})}_{\text{up-parabola flow }(+)}\end{array}\right\} \text{for } \alpha_{j_1j_2}>0,$$

$$\underbrace{(\text{nF},\text{UP})}_{\text{up-parabola flow }(-)} \rightleftharpoons \underbrace{(\text{DP},\text{SO})}_{\text{switching bifurcation}} \rightleftharpoons \left\{\begin{array}{c}\underbrace{(\text{DP},\text{nF})}_{\text{down-parabola flow }(-)}\\ \underbrace{(\text{pF},\text{DP})}_{\text{down-parabola flow }(+)}\\ \underbrace{(\text{DP},\text{pF})}_{\text{down-parabola flow }(+)}\end{array}\right\} \text{for } \alpha_{j_1j_2}<0. \tag{3.75}$$

3.2 Proof of Theorem 3.1

(i) From the local analysis and differential equation, the proof of results can be proved.
(ii) Consider

$$\dot{x}_{j_2} = a_{j_2 j_2 0}(y_{j_1} - a_{j_2 j_2 1}),$$
$$\dot{x}_{j_1} = a_{j_1 j_1 0}(y_{j_1} - a_{j_1 j_1 1})^2.$$

The corresponding dynamical system in the y_{j_1}-direction is

$$\dot{y}_{j_1} = \alpha_{j_1 j_1} a_{j_1 j_1 0}\left[(y_{j_1} - a_{j_1 j_1 1})^2 + c_{j_1 j_1}(y_{j_1} - a_{j_1 j_1 1}) + c_{j_1 j_1}(a_{j_1 j_1 1} - a_{j_2 j_2 1})\right]$$

and a dynamical system in the $y_{j_1}^{\perp}$-direction is

$$\dot{y}_{j_1}^{\perp} = \beta_{j_1 j_1} a_{j_1 j_1 0}(y_{j_1} - a_{j_1 j_1 1})^2 + \beta_{j_1 j_2} a_{j_2 j_2 0}(y_{j_1} - a_{j_2 j_2 1})$$

with

$$c_{j_1 j_1} = \frac{\alpha_{j_1 j_2} a_{j_2 j_2 0}}{\alpha_{j_1 j_1} a_{j_1 j_1 0}}, c_{j_1 j_1 1} = \frac{\beta_{j_1 j_2} a_{j_2 j_2 0}}{\beta_{j_1 j_1} a_{j_1 j_1 0}}.$$

Let

$$\Delta = (c_{j_1 j_1})^2 - 4 c_{j_1 j_1}(a_{j_2 j_2 1} - a_{j_1 j_1 1}).$$

Thus, there are three cases of equilibrium states in the y_{j_1}- direction.

- For $\Delta < 0$, there is no any equilibrium on the y_{j_1}- direction.
- For $\Delta = 0$, there is an equilibrium on the y_{j_1} - direction for the one-dimensional lower-saddle or upper-saddle flows.
- For $\Delta > 0$, there are two equilibriums on the y_{j_1} - direction for the one-dimensional sink and source flows.

(ii$_1$) For $y_{j_1}^* = a_{j_1 j_1 1}$ with $a_{j_1 j_1 1} \neq a_{j_2 j_2 1}$ for the x_{j_1}-direction,

$$\frac{dx_{j_1}}{dx_{j_2}}\bigg|_{y_{j_1}^* = a_{j_1 j_1 1}} = \frac{a_{j_1 j_1 0}(y_{j_1} - a_{j_1 j_1 1})^2}{a_{j_2 j_2 0}(y_{j_1} - a_{j_2 j_2 1})}\bigg|_{y_{j_1}^* = a_{j_1 j_1 1}} = 0,$$

and

$$\frac{d^2x_{j_1}}{dx_{j_2}^2}\Big|_{y_{j_1}^*=a_{j_1j_11}}=0.$$

If

$$\frac{d^3x_{j_1}}{dx_{j_2}^3}\Big|_{y_{j_1}^*=a_{j_1j_11}}=\frac{2a_{j_1j_10}\alpha_{j_1j_2}\alpha_{j_1j_2}}{a_{j_2j_20}(a_{j_1j_11}-a_{j_2j_21})}>0,$$

there is an increasing-inflection flow at $y_{j_1}^*=a_{j_1j_11}$ for the x_{j_1}-direction. If

$$\frac{d^3x_{j_1}}{dx_{j_2}^3}\Big|_{y_{j_1}^*=a_{j_1j_11}}=\frac{2a_{j_1j_10}\alpha_{j_1j_2}\alpha_{j_1j_2}}{a_{j_2j_20}(a_{j_1j_11}-a_{j_2j_21})}<0,$$

there is a decreasing-inflection flow at $y_{j_1}^*=a_{j_1j_11}$ for the x_{j_1}-direction. Let

$$\dot{x}_{j_2}^*=a_{j_2j_20}(a_{j_1j_11}-a_{j_2j_21}).$$

For $\dot{x}_{j_2}^*>0$ and $\dot{x}_{j_2}^*<0$, the flow at $y_{j_1}^*=a_{j_1j_11}$ for the x_{j_2}-direction is positive and negative accordingly.

Therefore, the equilibrium of $y_{j_1}^*=a_{j_1j_11}$ for the x_{j_1}-direction possesses the following properties with $\dot{x}_{j_2}^*=a_{j_2j_20}(a_{j_1j_11}-a_{j_2j_21})$ as in Eqs. (3.20)–(3.23):

- For $\alpha_{j_1j_2}^2a_{j_1j_10}>0$ and $a_{j_2j_20}(a_{j_1j_11}-a_{j_2j_21})>0$,

$$(a_{j_1j_11},\dot{x}_{j_2}^*)=\underbrace{(\text{II, pF})}_{\text{increasing-inflection flow }(+)}.$$

- For $\alpha_{j_1j_2}^2a_{j_1j_10}<0$ and $a_{j_2j_20}(a_{j_1j_11}-a_{j_2j_21})>0$,

$$(a_{j_1j_11},\dot{x}_{j_2}^*)=\underbrace{(\text{DI, pF})}_{\text{decreasing-inflection flow }(+)}.$$

- For $\alpha_{j_1j_2}^2a_{j_1j_10}>0$ and $a_{j_2j_20}(a_{j_1j_11}-a_{j_2j_21})<0$,

$$(a_{j_1j_11},\dot{x}_{j_2}^*)=\underbrace{(\text{DI, nF})}_{\text{decreasing-inflection flow }(-)}.$$

- For $\alpha_{j_1j_2}^2a_{j_1j_10}<0$ and $a_{j_2j_20}(a_{j_1j_11}-a_{j_2j_21})<0$,

$$(a_{j_1j_1 1}, \dot{x}^*_{j_2}) = \underbrace{(\text{II}, \text{nF})}_{\text{increasing-inflection flow } (-)}.$$

(ii$_2$) For $y^*_{j_1} = a_{j_2j_2 1}$ with $a_{j_1j_1 1} \neq a_{j_2j_2 1}$ for the x_{j_1}-direction,

$$\left.\frac{dx_{j_2}}{dx_{j_1}}\right|_{y^*_{j_2}=a_{j_2j_2 1}} = \left.\frac{a_{j_2j_2 0}(y_{j_1} - a_{j_2j_2 1})}{a_{j_1j_1 0}(y_{j_1} - a_{j_1j_1 1})^2}\right|_{y^*_{j_2}=a_{j_2j_2 1}} = 0.$$

If

$$\left.\frac{d^2x_{j_2}}{dx^2_{j_1}}\right|_{y^*_{j_2}=a_{j_2j_2 1}} = \frac{a_{j_2j_2 0}\alpha_{j_1j_1}}{a_{j_1j_1 0}(a_{j_2j_2 1} - a_{j_1j_1 1})^2} > 0,$$

there is an up-parabola flow at $y^*_{j_2} = a_{j_2j_2 1}$ for the x_{j_2}-direction. If

$$\left.\frac{d^2x_{j_2}}{dx^2_{j_1}}\right|_{y^*_{j_2}=a_{j_2j_2 1}} = \frac{a_{j_2j_2 0}\alpha_{j_1j_1}}{a_{j_1j_1 0}(a_{j_2j_2 1} - a_{j_1j_1 1})^2} < 0,$$

there is a down-parabola flow at $y^*_{j_2} = a_{j_2j_2 1}$ for the x_{j_2}-direction. Let

$$\dot{x}^*_{j_1} = a_{j_1j_1 0}(a_{j_2j_2 1} - a_{j_1j_1 1})^2.$$

For $\dot{x}^*_{j_1} > 0$ and $\dot{x}^*_{j_1} < 0$, the flow at $y^*_{j_1} = a_{j_1j_1 1}$ for the x_{j_1}-direction is positive and negative, respectively.

Therefore, the equilibrium of $y^*_{j_1} = a_{j_2j_2 1}$ for the x_{j_2}-direction possesses the following properties with $\dot{x}^*_{j_1} = a_{j_1j_1 0}(a_{j_2j_2 1} - a_{j_1j_1 1})^2$ as in Eqs. (3.24)–(3.27):

- For $a_{j_1j_1 0}(a_{j_2j_2 1} - a_{j_1j_1 1})^2 > 0$ and $\alpha_{j_1j_1}a_{j_2j_2 0} > 0$,

$$(\dot{x}^*_{j_1}, a_{j_2j_2 1}) = \underbrace{(\text{pF}, \text{UP})}_{\text{up-parabola flow } (+)}.$$

- For $a_{j_1j_1 0}(a_{j_2j_2 1} - a_{j_1j_1 1})^2 < 0$ and $\alpha_{j_1j_1}a_{j_2j_2 0} > 0$,

$$(\dot{x}^*_{j_1}, a_{j_2j_2 1}) = \underbrace{(\text{nF}, \text{DP})}_{\text{down-parabola flow } (-)}.$$

- For $a_{j_1j_10}(a_{j_2j_21}-a_{j_1j_11})^2>0$ and $\alpha_{j_1j_1}a_{j_2j_20}<0$,

$$(\dot{x}^*_{j_1}, a_{j_2j_21}) = \underbrace{(\text{pF}, \text{DP})}_{\text{down-parabola flow }(+)}.$$

- For $a_{j_1j_10}(a_{j_2j_21}-a_{j_1j_11})^2<0$ and $\alpha_{j_1j_1}a_{j_2j_20}<0$,

$$(\dot{x}^*_{j_1}, a_{j_2j_21}) = \underbrace{(\text{nF}, \text{UP})}_{\text{up-parabola flow }(-)}.$$

(ii$_3$) For $a_{j_2j_21}=a_{j_1j_11}$, consider

$$\begin{aligned}\dot{x}_{j_2} &= a_{j_2j_20}(y_{j_1}-a_{j_1j_11}),\\ \dot{x}_{j_1} &= a_{j_1j_10}(y_{j_1}-a_{j_1j_11})^2.\end{aligned}$$

The dynamical system in the y_{j_1}-direction is constructed by

$$\dot{y}_{j_1} = \alpha_{j_1j_1}a_{j_1j_10}(y_{j_1}-a_{j_1j_11})(y_{j_1}-a_{j_1j_11}+c_{j_1j_1})$$

and there are two equilibriums on the y_{j_1}- direction for the one-dimensional sink and source flows, and a dynamical system in the $y^{\perp}_{j_1}$-direction is developed by

$$\dot{y}^{\perp}_{j_1} = \beta_{j_1j_1}a_{j_1j_10}(y_{j_1}-a_{j_1j_11})^2+\beta_{j_1j_2}a_{j_2j_20}(y_{j_1}-a_{j_1j_11}).$$

For $y^*_{j_1}=a_{j_1j_11}$, we have

$$\left.\frac{dx_{j_1}}{dx_{j_2}}\right|_{y^*_{j_1}=a_{j_1j_11}} = \frac{a_{j_1j_10}(y_{j_1}-a_{j_1j_11})}{a_{j_2j_20}}.$$

If

$$\left.\frac{d^2x_{j_1}}{dx^2_{j_2}}\right|_{y^*_{j_1}=a_{j_1j_11}} = \frac{\alpha_{j_1j_2}a_{j_1j_10}}{a_{j_2j_20}} = \frac{\alpha^2_{j_1j_2}a_{j_1j_10}}{\alpha_{j_1j_2}a_{j_2j_20}} > 0,$$

there is an up-parabola flow at $y^*_{j_1}=a_{j_1j_11}$ for x_{j_1}-direction. If

$$\left.\frac{d^2x_{j_1}}{dx^2_{j_2}}\right|_{y^*_{j_1}=a_{j_1j_11}} = \frac{\alpha_{j_1j_2}a_{j_1j_10}}{a_{j_2j_20}} = \frac{\alpha^2_{j_1j_2}a_{j_1j_10}}{\alpha_{j_1j_2}a_{j_2j_20}} < 0,$$

there is a down-parabola flow at $y^*_{j_1}=a_{j_1j_11}$ for x_{j_1}-direction. The variational equation is

$$\Delta\dot{y}_{j_1} = \alpha_{j_1 j_1} a_{j_1 j_1 0} c_{j_1 j_1} \Delta y_{j_1} = \alpha_{j_1 j_2} a_{j_2 j_2 0} \Delta y_{j_1}.$$

where

$$G_{j_1 j_1 0} = \alpha_{j_1 j_2} a_{j_2 j_2 0}.$$

For $\alpha_{j_1 j_2} a_{j_2 j_2 0} > 0$, there is a parabola source flow, and for $\alpha_{j_1 j_2} a_{j_2 j_2 0} < 0$, there is a parabola sink flow.

Therefore, the infinite-equilibriums of $y_{j_1}^* = a_{j_1 j_1 1}$ for the x_{j_1}-direction and the x_{j_2}-direction possess the following properties as in Eqs. (3.31)–(3.34).

- For $\alpha_{j_1 j_2}^2 a_{j_1 j_1 0} > 0$ and $\alpha_{j_1 j_2} a_{j_2 j_2 0} > 0$,

$$(a_{j_1 j_1 1}, a_{j_2 j_2 1}) = \underbrace{(\text{UP}, \text{SO})}_{\text{up-parabola source}}.$$

- For $\alpha_{j_1 j_2}^2 a_{j_1 j_1 0} < 0$ and $\alpha_{j_1 j_2} a_{j_2 j_2 0} > 0$,

$$(a_{j_1 j_1 1}, a_{j_2 j_2 1}) = \underbrace{(\text{DP}, \text{SO})}_{\text{down-parabola source}}.$$

- For $\alpha_{j_1 j_2}^2 a_{j_1 j_1 0} > 0$ and $\alpha_{j_1 j_2} a_{j_2 j_2 0} < 0$,

$$(a_{j_1 j_1 1}, a_{j_2 j_2 1}) = \underbrace{(\text{DP}, \text{SI})}_{\text{down-parabola sink}}.$$

- For $\alpha_{j_1 j_2}^2 a_{j_1 j_1 0} < 0$ and $\alpha_{j_1 j_2} a_{j_2 j_2 0} < 0$,

$$(a_{j_1 j_1 1}, a_{j_2 j_2 1}) = \underbrace{(\text{UP}, \text{SI})}_{\text{up-parabola sink}}.$$

(ii$_4$) For $a_{j_2 j_2 1} = a_{j_1 j_1 1}$, from case (ii$_1$)-(ii$_3$), it is obtained the infinite-equilibriums of $y_{j_1}^* = a_{j_1 j_1 1}$ is a switching bifurcation and four cases are obtained as in Eqs. (3.35)–(3.38).

- *For* $a_{j_1 j_1 0} > 0$ and $a_{j_2 j_2 0} > 0$,

$$\left\{\begin{array}{c}\underbrace{(\mathrm{pF},\mathrm{UP})}_{\text{up-parabola flow }(+)}\\\underbrace{(\mathrm{II},\mathrm{pF})}_{\text{increasing-inflection flow }(-)}\end{array}\right\}\rightleftharpoons\underbrace{(\mathrm{UP},\mathrm{SO})}_{\text{switching bifurcation}}\rightleftharpoons\left\{\begin{array}{c}\underbrace{(\mathrm{DI},\mathrm{nF})}_{\text{decreasing-inflection flow }(-)}\\\underbrace{(\mathrm{pF},\mathrm{UP})}_{\text{up-parabola flow }(+)}\end{array}\right\}$$

for $\alpha_{j_1 j_2} > 0$;

$$\left\{\begin{array}{c}\underbrace{(\mathrm{pF},\mathrm{UP})}_{\text{up-parabola flow }(+)}\\\underbrace{(\mathrm{II},\mathrm{pF})}_{\text{increasing-inflection flow }(+)}\end{array}\right\}\rightleftharpoons\underbrace{(\mathrm{DP},\mathrm{SI})}_{\text{switching-bifurcation}}\rightleftharpoons\left\{\begin{array}{c}\underbrace{(\mathrm{DI},\mathrm{nF})}_{\text{decreasing-inflection flow }(-)}\\\underbrace{(\mathrm{pF},\mathrm{UP})}_{\text{up-parabola flow }(+)}\end{array}\right\}$$

for $\alpha_{j_1 j_2} < 0$.

- *For* $a_{j_1 j_1 0} < 0$ and $a_{j_2 j_2 0} > 0$,

$$\left\{\begin{array}{c}\underbrace{(\mathrm{nF},\mathrm{DP})}_{\text{down-parabola flow }(-)}\\\underbrace{(\mathrm{DI},\mathrm{pF})}_{\text{decreasing-inflection flow }(+)}\end{array}\right\}\rightleftharpoons\underbrace{(\mathrm{DP},\mathrm{SO})}_{\text{switching-bifurcation}}\rightleftharpoons\left\{\begin{array}{c}\underbrace{(\mathrm{II},\mathrm{nF})}_{\text{increasing-inflection flow }(-)}\\\underbrace{(\mathrm{nF},\mathrm{DP})}_{\text{down-parabola flow }(-)}\end{array}\right\}$$

for $\alpha_{j_1 j_2} > 0$;

$$\left\{\begin{array}{c}\underbrace{(\mathrm{nF},\mathrm{DP})}_{\text{down-parabola flow }(-)}\\\underbrace{(\mathrm{DI},\mathrm{pF})}_{\text{decreasing-inflection flow }(+)}\end{array}\right\}\rightleftharpoons\underbrace{(\mathrm{UP},\mathrm{SI})}_{\text{switching-bifurcation}}\rightleftharpoons\left\{\begin{array}{c}\underbrace{(\mathrm{II},\mathrm{nF})}_{\text{increasing-inflection flow }(-)}\\\underbrace{(\mathrm{nF},\mathrm{DP})}_{\text{down-parabola flow }(-)}\end{array}\right\}$$

for $\alpha_{j_1 j_2} > 0$.

- *For* $a_{j_1 j_1 0} > 0$ and $a_{j_2 j_2 0} < 0$,

$$\left\{\begin{array}{c}\underbrace{(\mathrm{pF},\mathrm{DP})}_{\text{down-parabola flow }(+)}\\ \underbrace{(\mathrm{DI},\mathrm{nF})}_{\text{decreasing-inflection flow }(-)}\end{array}\right\} \rightleftharpoons \underbrace{(\mathrm{DP},\mathrm{SI})}_{\text{switching-bifurcation}} \rightleftharpoons \left\{\begin{array}{c}\underbrace{(\mathrm{II},\mathrm{pF})}_{\text{increasing-inflection flow }(+)}\\ \underbrace{(\mathrm{pF},\mathrm{DP})}_{\text{down-parabola flow }(+)}\end{array}\right\}$$

for $\alpha_{j_1 j_2} > 0$;

$$\left\{\begin{array}{c}\underbrace{(\mathrm{pF},\mathrm{DP})}_{\text{down-parabola flow }(+)}\\ \underbrace{(\mathrm{DI},\mathrm{nF})}_{\text{decreasing-inflection flow }(-)}\end{array}\right\} \rightleftharpoons \underbrace{(\mathrm{UP},\mathrm{SO})}_{\text{switching-bifucations}} \rightleftharpoons \left\{\begin{array}{c}\underbrace{(\mathrm{II},\mathrm{pF})}_{\text{increasing-inflection flow }(+)}\\ \underbrace{(\mathrm{pF},\mathrm{DP})}_{\text{down-parabola flow }(+)}\end{array}\right\}$$

for $\alpha_{j_1 j_2} < 0$.

- *For* $a_{j_1 j_1 0} < 0$ and $a_{j_2 j_2 0} < 0$,

$$\left\{\begin{array}{c}\underbrace{(\mathrm{nF},\mathrm{UP})}_{\text{up-parabola flow }(-)}\\ \underbrace{(\mathrm{II},\mathrm{nF})}_{\text{increasing-inflection flow }(-)}\end{array}\right\} \rightleftharpoons \underbrace{(\mathrm{UP},\mathrm{SI})}_{\text{switching-bifurcation}} \rightleftharpoons \left\{\begin{array}{c}\underbrace{(\mathrm{DI},\mathrm{pF})}_{\text{decreasing-inflection flow }(+)}\\ \underbrace{(\mathrm{nF},\mathrm{UP})}_{\text{up-parabola flow }(-)}\end{array}\right\}$$

for $\alpha_{j_1 j_2} > 0$;

$$\left\{\begin{array}{c}\underbrace{(\mathrm{nF},\mathrm{UP})}_{\text{up-parabola flow }(-)}\\ \underbrace{(\mathrm{II},\mathrm{nF})}_{\text{increasing-inflection flow }(-)}\end{array}\right\} \rightleftharpoons \underbrace{(\mathrm{DP},\mathrm{SO})}_{\text{switching-bifurcation}} \rightleftharpoons \left\{\begin{array}{c}\underbrace{(\mathrm{DI},\mathrm{pF})}_{\text{decreasing-inflection flow }(+)}\\ \underbrace{(\mathrm{nF},\mathrm{UP})}_{\text{up-parabola flow }(-)}\end{array}\right\}$$

for $\alpha_{j_1 j_2} < 0$.

(iii) Consider a dynamical system as

$$\begin{aligned}\dot{x}_{j_2} &= a_{j_2 j_2 0}(y_{j_1} - a_{j_2 j_2 1}),\\ \dot{x}_{j_1} &= a_{j_1 j_1 0}(y_{j_1} - a_{j_1 j_1 1})(y_{j_1} - a_{j_1 j_1 2}).\end{aligned}$$

The corresponding dynamical system in the y_{j_1}-direction is constructed by

$$\dot{y}_{j_1} = \alpha_{j_1 j_1} a_{j_1 j_1 0}\left[y_{j_1}^2 - (a_{j_1 j_1 1} + a_{j_1 j_1 2} - c_{j_1 j_1})y_{j_1} + (a_{j_1 j_1 1} a_{j_1 j_1 2} - c_{j_1 j_1} a_{j_2 j_2 1})\right],$$

and a dynamical system in the $y_{j_1}^{\perp}$-direction is constructed by

$$\dot{y}_{j_1}^{\perp} = \beta_{j_1 j_1} a_{j_1 j_1 0}(y_{j_1} - a_{j_1 j_1 1})(y_{j_1} - a_{j_1 j_1 2}) + \beta_{j_1 j_2} a_{j_2 j_2 0}(y_{j_1} - a_{j_2 j_2 1})$$

and

$$\Delta = (a_{j_1 j_1 1} + a_{j_1 j_1 2} - c_{j_1 j_1})^2 - 4(a_{j_1 j_1 1} a_{j_1 j_1 2} - c_{j_1 j_1} a_{j_2 j_2 1}).$$

There are three cases of the equilibriums on the y_{j_1}-direction.

- For $\Delta < 0$, there is no equilibrium on the y_{j_1}-direction.
- For $\Delta = 0$, there is an equilibrium on the y_{j_1}-direction for the one-dimensional lower-saddle or upper-saddle flows.
- For $\Delta > 0$, there are two equilibriums on the y_{j_1}-direction for the one-dimensional sink and source flows.

(iii$_1$) For $y_{j_1}^* = a_{j_1 j_1 s}$ with $y_{j_1} \neq a_{j_2 j_2 1}$,

$$\frac{dx_{j_1}}{dx_{j_2}}\bigg|_{y_{j_1}^* = a_{j_1 j_1 s}} = \frac{a_{j_1 j_1 0}(y_{j_1} - a_{j_1 j_1 s})(y_{j_1} - a_{j_1 j_1 l})}{a_{j_2 j_2 0}(y_{j_1} - a_{j_2 j_2 1})}\bigg|_{y_{j_1}^* = a_{j_1 j_1 s}} = 0.$$

If

$$\frac{d^2 x_{j_1}}{dx_{j_2}^2}\bigg|_{y_{j_1}^* = a_{j_1 j_1 s}} = \frac{a_{j_1 j_1 0}\alpha_{j_1 j_2}(a_{j_1 j_1 s} - a_{j_1 j_1 l})}{a_{j_2 j_2 0}(a_{j_1 j_1 s} - a_{j_2 j_2 1})} > 0,$$

there is an up-parabola flow near $y_{j_1}^* = a_{j_1 j_1 s}$. If

$$\frac{d^2 x_{j_1}}{dx_{j_2}^2}\bigg|_{y_{j_1}^* = a_{j_1 j_1 s}} = \frac{a_{j_1 j_1 0}\alpha_{j_1 j_2}(a_{j_1 j_1 s} - a_{j_1 j_1 l})}{a_{j_2 j_2 0}(a_{j_1 j_1 s} - a_{j_2 j_2 1})} < 0,$$

there is a down-parabola flow near $y_{j_1}^* = a_{j_1 j_1 s}$. Let

$$\dot{x}_{j_2}^* = a_{j_2 j_2 0}(a_{j_1 j_1 s} - a_{j_2 j_2 1}).$$

For $\dot{x}_{j_2}^* > 0$ and $\dot{x}_{j_2}^* < 0$, the flow at $y_{j_1}^* = a_{j_1 j_1 s}$ in the x_{j_2}-direction is positive and negative, respectively.

Therefore, the equilibrium of $y_{j_1}^* = a_{j_1 j_1 s}$ ($s, l \in \{1,2\}, s \neq l$) for the x_{j_1}-direction possesses the following properties with $\dot{x}_{j_2}^* = a_{j_2 j_2 0}(a_{j_1 j_1 s} - a_{j_2 j_2 1})$ as in Eqs. (3.46)–(3.49):

- For $a_{j_1j_10}\alpha_{j_1j_2}(a_{j_1j_1s}-a_{j_1j_1l})>0$ and $a_{j_2j_20}(a_{j_1j_1s}-a_{j_2j_21})>0$,

$$(a_{j_1j_1s}, \dot{x}^*_{j_2}) = \underbrace{(\mathrm{UP}, \mathrm{pF})}_{\text{up-parabola flow }(+)}.$$

- For $a_{j_1j_10}\alpha_{j_1j_2}(a_{j_1j_1s}-a_{j_1j_1l})<0$ and $a_{j_2j_20}(a_{j_1j_1s}-a_{j_2j_21})>0$,

$$(a_{j_1j_1s}, \dot{x}^*_{j_2}) = \underbrace{(\mathrm{DP}, \mathrm{pF})}_{\text{down-parabola flow }(+)}.$$

- For $a_{j_1j_10}\alpha_{j_1j_2}(a_{j_1j_1s}-a_{j_1j_1l})>0$ and $a_{j_2j_20}(a_{j_1j_1s}-a_{j_2j_21})<0$,

$$(a_{j_1j_1s}, \dot{x}^*_{j_2}) = \underbrace{(\mathrm{DP}, \mathrm{nF})}_{\text{down-parabola flow }(-)}.$$

- For $a_{j_1j_10}\alpha_{j_1j_2}(a_{j_1j_1s}-a_{j_1j_1l})<0$ and $a_{j_2j_20}(a_{j_1j_1s}-a_{j_2j_21})<0$,

$$(a_{j_1j_1s}, \dot{x}^*_{j_2}) = \underbrace{(\mathrm{UP}, \mathrm{nF})}_{\text{up-parabola flow }(-)}.$$

(iii$_2$) For $y^*_{j_1} = a_{j_2j_21}$ with $y_{j_1} \neq a_{j_1j_11}, a_{j_1j_12}$,

$$\frac{dx_{j_2}}{dx_{j_1}}\Big|_{y^*_{j_1}=a_{j_2j_21}} = \frac{a_{j_2j_20}(y_{j_1}-a_{j_2j_21})}{a_{j_1j_10}(y_{j_1}-a_{j_1j_11})(y_{j_1}-a_{j_1j_12})}\Big|_{y^*_{j_1}=a_{j_2j_21}} = 0.$$

If

$$\frac{d^2x_{j_2}}{dx^2_{j_1}}\Big|_{y^*_{j_1}=a_{j_2j_21}} = \frac{a_{j_2j_20}\alpha_{j_1j_1}}{a_{j_1j_10}(a_{j_2j_21}-a_{j_1j_11})(a_{j_2j_21}-a_{j_1j_12})} > 0,$$

there is an up-parabola flow at $y^*_{j_1} = a_{j_2j_21}$ in the y_{j_2} -direction. If

$$\frac{d^2x_{j_2}}{dx^2_{j_1}}\Big|_{y^*_{j_1}=a_{j_2j_21}} = \frac{a_{j_2j_20}\alpha_{j_1j_1}}{a_{j_1j_10}(a_{j_2j_21}-a_{j_1j_11})(a_{j_2j_21}-a_{j_1j_12})} < 0,$$

there is a down-parabola flow at $y^*_{j_1} = a_{j_2j_21}$ in the y_{j_2} -direction. Let

$$\dot{x}^*_{j_1} = a_{j_1j_10}(a_{j_2j_21}-a_{j_1j_11})(a_{j_2j_21}-a_{j_1j_12}).$$

For $\dot{x}^*_{j_1} > 0$ and $\dot{x}^*_{j_1} < 0$, the flow at $y^*_{j_1} = a_{j_1j_11}$ in the x_{j_2}-direction is positive and negative, respectively.

Therefore, the equilibrium of $y^*_{j_1} = a_{j_2j_2 1}$ for the x_{j_2}-direction possesses the following properties with $\dot{x}^*_{j_1} = a_{j_1j_1 0}\prod_{s=1}^{2}(a_{j_2j_2 1} - a_{j_1j_1 s})$ as in Eqs. (3.50)–(3.53):

- For $a_{j_1j_1 0}\prod_{s=1}^{2}(a_{j_2j_2 1} - a_{j_1j_1 s}) > 0$ and $a_{j_2j_2 0}\alpha_{j_1j_1} > 0$,

$$(\dot{x}^*_{j_1}, a_{j_2j_2 1}) = \underbrace{(\text{pF}, \text{UP})}_{\text{up-parabola flow } (+)}.$$

- For $a_{j_1j_1 0}\prod_{s=1}^{2}(a_{j_2j_2 1} - a_{j_1j_1 s}) < 0$ and $a_{j_2j_2 0}\alpha_{j_1j_1} > 0$,

$$(\dot{x}^*_{j_1}, a_{j_2j_2 1}) = \underbrace{(\text{nF}, \text{DP})}_{\text{down-parabola flow } (-)}.$$

- For $a_{j_1j_1 0}\prod_{s=1}^{2}(a_{j_2j_2 1} - a_{j_1j_1 s}) > 0$ and $a_{j_2j_2 0}\alpha_{j_1j_1} < 0$,

$$(\dot{x}^*_{j_1}, a_{j_2j_2 1}) = \underbrace{(\text{pF}, \text{DP})}_{\text{down-parabola flow } (+)}.$$

- For $a_{j_1j_1 0}\prod_{s=1}^{2}(a_{j_2j_2 1} - a_{j_1j_1 s}) < 0$ and $a_{j_2j_2 0}\alpha_{j_1j_1} < 0$,

$$(\dot{x}^*_{j_1}, a_{j_2j_2 1}) = \underbrace{(\text{nF}, \text{UP})}_{\text{up-parabola flow } (-)}.$$

(iv) For $a_{j_2j_2 1} = a_{j_1j_1 s}$ ($s, l \in \{1, 2\}$, $s \neq l$), consider a dynamical system as

$$\begin{aligned}\dot{x}_{j_2} &= a_{j_2j_2 0}(y_{j_1} - a_{j_2j_2 1}),\\ \dot{x}_{j_1} &= a_{j_1j_1 0}(y_{j_1} - a_{j_1j_1 s})(y_{j_1} - a_{j_1j_1 l}).\end{aligned}$$

The corresponding dynamical system in the y_{j_1}-direction is

$$\dot{y}_{j_1} = \alpha_{j_1j_1} a_{j_1j_1 0}(y_{j_1} - a_{j_1j_1 s})(y_{j_1} - a_{j_1j_1 l} + c_{j_1j_1}),$$

and a dynamical system in the $y^{\perp}_{j_1}$-direction is

$$\dot{y}^{\perp}_{j_1} = \beta_{j_1j_1} a_{j_1j_1 0}(y_{j_1} - a_{j_1j_1 s})(y_{j_1} - a_{j_1j_1 2} + c_{j_1j_1 1}),$$

with

$$c_{j_1j_1} = \frac{\alpha_{j_1j_2} a_{j_2j_2 0}}{\alpha_{j_1j_1} a_{j_1j_1 0}}, c_{j_1j_1 1} = \frac{\beta_{j_1j_2} a_{j_2j_2 0}}{\beta_{j_1j_1} a_{j_1j_1 0}}.$$

Thus, the equilibrium of $y_{j_1} = a_{j_1j_1 s}$ is an infinite-equilibrium.

(iv$_1$) For $y^*_{j_1} = a_{j_1 j_1 s} = a_{j_2 j_2 1} (s,\ l \in \{1, 2\},\ s \neq l)$, if

$$\left.\frac{dx_{j_1}}{dx_{j_2}}\right|_{y^*_{j_1} = a_{j_1 j_1 s}} = \frac{\alpha_{j_1 j_1} a_{j_1 j_1 0}(a_{j_1 j_1 s} - a_{j_1 j_1 l})}{\alpha_{j_1 j_1} a_{j_2 j_2 0}} > 0,$$

there is an increasing-inflection flow for $y^*_{j_1} = a_{j_1 j_1 s} = a_{j_2 j_2 1}$; and if

$$\left.\frac{dx_{j_1}}{dx_{j_2}}\right|_{y^*_{j_1} = a_{j_1 j_1 s}} = \frac{\alpha_{j_1 j_1} a_{j_1 j_1 0}(a_{j_1 j_1 s} - a_{j_1 j_1 l})}{\alpha_{j_1 j_1} a_{j_2 j_2 0}} < 0,$$

there is a decreasing-inflection flow for $y^*_{j_1} = a_{j_1 j_1 s} = a_{j_2 j_2 1}$.

In the vicinity of $y^*_{j_1} = a_{j_1 j_1 s} = a_{j_2 j_2 1}$, we have

$$\Delta \dot{y}_{j_1} = \left[\alpha_{j_1 j_1} a_{j_1 j_1 0}(a_{j_1 j_1 s} - a_{j_1 j_1 l}) + \alpha_{j_1 j_2} a_{j_2 j_2 0}\right] \Delta y_{j_1} = G_{j_1 j_1 0} \Delta y_{j_1},$$

with

$$G_{j_1 j_1 0} = \alpha_{j_1 j_1} a_{j_1 j_1 0}(a_{j_1 j_1 s} - a_{j_1 j_1 l}) + \alpha_{j_1 j_2} a_{j_2 j_2 0}.$$

The inflection flows are source and sink for $G_{j_1 j_1 0} > 0$ and $G_{j_1 j_1 0} < 0$.

Therefore, the infinite-equilibriums of $y^*_{j_1} = a_{j_1 j_1 s}\ (s,\ l \in \{1, 2\},\ s \neq l)$ for the x_{j_1}-direction and $y^*_{j_1} = a_{j_2 j_2 1}$ for the x_{j_2}-direction possesses the following properties as in Eqs. (3.58)–(3.61):

- For $\alpha_{j_1 j_1} a_{j_1 j_1 0}(a_{j_1 j_1 s} - a_{j_1 j_1 l})/\alpha_{j_1 j_1} a_{j_2 j_2 0} > 0$ and $G_{j_1 j_1 0} > 0$,

$$(a_{11s}, a_{j_2 j_2 1}) = \underbrace{(\mathrm{II}, \mathrm{SO})}_{\text{increasing-inflection source}}.$$

- For $\alpha_{j_1 j_1} a_{j_1 j_1 0}(a_{j_1 j_1 s} - a_{j_1 j_1 l})/\alpha_{j_1 j_1} a_{j_2 j_2 0} > 0$ and $G_{j_1 j_1 0} < 0$,

$$(a_{11s}, a_{j_2 j_2 1}) = \underbrace{(\mathrm{II}, \mathrm{SI})}_{\text{increasing-inflection sink}}.$$

- For $\alpha_{j_1 j_1} a_{j_1 j_1 0}(a_{j_1 j_1 s} - a_{j_1 j_1 l})/\alpha_{j_1 j_1} a_{j_2 j_2 0} < 0$ and $G_{j_1 j_1 0} > 0$,

$$(a_{11s}, a_{j_2 j_2 1}) = \underbrace{(\mathrm{DI}, \mathrm{SO})}_{\text{decreasing-inflection source}}.$$

- For $\alpha_{j_1 j_1} a_{j_1 j_1 0}(a_{j_1 j_1 s} - a_{j_1 j_1 l})/\alpha_{j_1 j_1} a_{j_2 j_2 0} < 0$ and $G_{j_1 j_1 0} < 0$,

$$(a_{11s}, a_{j_2 j_2 1}) = \underbrace{(\mathrm{DI}, \mathrm{SI})}_{\text{decreasing-inflection sink}}.$$

(iv$_2$) For $y_{j_1}^* = a_{j_1 j_1 l}(s, l \in \{1,2\}, l \neq s)$,

$$\frac{dx_{j_1}}{dx_{j_2}}\bigg|_{y_{j_1}^* = a_{j_1 j_1 l}} = \frac{a_{j_1 j_1 0}(y_{j_1} - a_{j_1 j_1 l})}{a_{j_2 j_2 0}}\bigg|_{y_{j_1}^* = a_{j_1 j_1 l}} = 0.$$

If

$$\frac{d^2 x_{j_1}}{dx_{j_2}^2}\bigg|_{y_{j_1}^* = a_{j_1 j_1 l}} = \frac{\alpha_{j_1 j_2} a_{j_1 j_1 0}}{a_{j_2 j_2 0}} = \frac{\alpha_{j_1 j_1}\alpha_{j_1 j_2} a_{j_1 j_1 0}(a_{j_1 j_1 l} - a_{j_1 j_1 s})}{\alpha_{j_1 j_1} a_{j_2 j_2 0}(a_{j_1 j_1 l} - a_{j_2 j_2 1})} > 0,$$

there is an up-parabola flow at $y_{j_1}^* = a_{j_1 j_1 l}$ in the x_{j_1}-direction. If

$$\frac{d^2 x_{j_1}}{dx_{j_2}^2}\bigg|_{y_{j_1}^* = a_{j_1 j_1 l}} = \frac{\alpha_{j_1 j_2} a_{j_1 j_1 0}}{a_{j_2 j_2 0}} = \frac{\alpha_{j_1 j_1}\alpha_{j_1 j_2} a_{j_1 j_1 0}(a_{j_1 j_1 l} - a_{j_1 j_1 s})}{\alpha_{j_1 j_1} a_{j_2 j_2 0}(a_{j_1 j_1 l} - a_{j_2 j_2 1})} < 0,$$

there is a down-parabola flow at $y_{j_1}^* = a_{j_1 j_1 l}$ in the x_{j_1}-direction. In the x_{j_2}-direction, the following relation holds,

$$\dot{x}_{j_2}^* = \alpha_{j_1 j_1} a_{j_2 j_2 0}(a_{j_1 j_1 l} - a_{j_2 j_2 1}).$$

For $\dot{x}_{j_2}^* > 0$ and $\dot{x}_{j_2}^* < 0$, the flows at $y_{j_1}^* = a_{j_1 j_1 l}$ are positive and negative, respectively.

Therefore, the equilibrium of $y_{j_1}^* = a_{j_1 j_1 l}$ $(l,\, s \in \{1,2\},\, l \neq s)$ for the x_{j_1}-direction with $\dot{x}_{j_2}^* = a_{j_2 j_2 0}(a_{j_1 j_1 l} - a_{j_2 j_2 1})$ possesses the following properties as in Eqs. (3.62)–(3.65):

- For $\alpha_{j_1 j_2} a_{j_1 j_1 0}(a_{j_1 j_1 l} - a_{j_1 j_1 s}) > 0$ and $a_{j_2 j_2 0}(a_{j_1 j_1 l} - a_{j_2 j_2 1}) > 0$,

$$(a_{j_1 j_1 l}, \dot{x}_{j_2}^*) = \underbrace{(\text{UP}, \text{pF})}_{\text{up-parabola flow }(+)}.$$

- For $\alpha_{j_1 j_2} a_{j_1 j_1 0}(a_{j_1 j_1 l} - a_{j_1 j_1 s}) < 0$ and $a_{j_2 j_2 0}(a_{j_1 j_1 l} - a_{j_2 j_2 1}) > 0$,

$$(a_{j_1 j_1 l}, \dot{x}_{j_2}^*) = \underbrace{(\text{DP}, \text{pF})}_{\text{down-parabola flow }(+)}.$$

- For $\alpha_{j_1 j_2} a_{j_1 j_1 0}(a_{j_1 j_1 l} - a_{j_1 j_1 s}) > 0$ and $a_{j_2 j_2 0}(a_{j_1 j_1 l} - a_{j_2 j_2 1}) < 0$,

$$(a_{j_1 j_1 l}, \dot{x}_{j_2}^*) = \underbrace{(\text{DP}, \text{nF})}_{\text{down-parabola flow }(-)}.$$

- For $\alpha_{j_1 j_2} a_{j_1 j_1 0}(a_{j_1 j_1 l} - a_{j_1 j_1 s}) < 0$ and $a_{j_2 j_2 0}(a_{j_1 j_1 l} - a_{j_2 j_2 1}) < 0$,

$$(a_{j_1 j_1 l}, \dot{x}_{j_2}^*) = \underbrace{(\text{UP}, \text{nF})}_{\text{up-parabola flow }(-)}.$$

(iv$_3$) For $y_{j_1}^* = a_{j_1j_1s} = a_{j_2j_21}(s,\ l \in \{1,2\},\ l \neq s)$, the infinite equilibrium of $\left(a_{j_1j_1l}, a_{j_2j_21}\right)$ is a switching bifurcations. From cases (iv$_1$) and (iv$_2$), the switching bifurcations have following four cases with $\alpha_{j_1j_1} > 0$ as in Eqs. (3.66)–(3.69).

- For $a_{j_1j_10}\left(a_{j_1j_1s} - a_{j_1j_1l}\right) > 0$ and $a_{j_2j_20} > 0$,

$$\left\{\begin{array}{c}\underbrace{(\mathrm{nF},\mathrm{DP})}_{\text{down-parabola flow }(-)}\\ \underbrace{(\mathrm{UP},\mathrm{pF})}_{\text{up-parabola flow }(+)}\end{array}\right\} \rightleftharpoons \underbrace{(\mathrm{II},\mathrm{SO})}_{\text{switching bifurcation}} \rightleftharpoons \left\{\begin{array}{c}\underbrace{(\mathrm{DP},\mathrm{nF})}_{\text{down-parabola flow }(-)}\\ \underbrace{(\mathrm{pF},\mathrm{UP})}_{\text{up-parabola flow }(+)}\end{array}\right\}$$

for $\alpha_{j_1j_2} > 0, G_{j_1j_10} > 0$;

$$\left\{\begin{array}{c}\underbrace{(\mathrm{nF},\mathrm{DP})}_{\text{down-parabola flow }(-)}\\ \underbrace{(\mathrm{DP},\mathrm{pF})}_{\text{down-parabola flow }(+)}\end{array}\right\} \rightleftharpoons \underbrace{(\mathrm{II},\mathrm{SO})}_{\text{switching bifurcation}} \rightleftharpoons \left\{\begin{array}{c}\underbrace{(\mathrm{UP},\mathrm{nF})}_{\text{up-parabola flow }(-)}\\ \underbrace{(\mathrm{pF},\mathrm{UP})}_{\text{up-parabola flow }(+)}\end{array}\right\}$$

for $\alpha_{j_1j_2} < 0, G_{j_1j_10} > 0$;

$$\left\{\begin{array}{c}\underbrace{(\mathrm{nF},\mathrm{DP})}_{\text{down-parabola flow }(-)}\\ \underbrace{(\mathrm{DP},\mathrm{pF})}_{\text{down-parabola flow }(+)}\end{array}\right\} \rightleftharpoons \underbrace{(\mathrm{II},\mathrm{SI})}_{\text{switching bifurcation}} \rightleftharpoons \left\{\begin{array}{c}\underbrace{(\mathrm{UP},\mathrm{nF})}_{\text{up-parabola flow }(-)}\\ \underbrace{(\mathrm{pF},\mathrm{UP})}_{\text{up-parabola flow }(+)}\end{array}\right\}$$

for $\alpha_{j_1j_2} < 0, G_{j_1j_10} < 0$.

- For $a_{j_1j_10}\left(a_{j_1j_1s} - a_{j_1j_1l}\right) < 0$ and $a_{j_2j_20} > 0$,

$$\left\{\begin{array}{c}\underbrace{(\mathrm{pF},\mathrm{UP})}_{\text{up-parabola flow }(+)}\\ \underbrace{(\mathrm{DP},\mathrm{pF})}_{\text{down-parabola flow }(+)}\end{array}\right\} \rightleftharpoons \underbrace{(\mathrm{DI},\mathrm{SI})}_{\text{switching bifurcation}} \rightleftharpoons \left\{\begin{array}{c}\underbrace{(\mathrm{UP},\mathrm{nF})}_{\text{up-parabola flow }(-)}\\ \underbrace{(\mathrm{nF},\mathrm{DP})}_{\text{down-parabola flow }(-)}\end{array}\right\}$$

for $\alpha_{j_1j_2} > 0, G_{j_1j_10} < 0$;

$$\left\{\begin{array}{c}\underbrace{(\mathrm{pF},\mathrm{UP})}_{\text{up-parabola flow }(+)}\\ \underbrace{(\mathrm{DP},\mathrm{pF})}_{\text{down-parabola flow }(+)}\end{array}\right\} \rightleftharpoons \underbrace{(\mathrm{DI},\mathrm{SO})}_{\text{switching bifurcation}} \rightleftharpoons \left\{\begin{array}{c}\underbrace{(\mathrm{UP},\mathrm{nF})}_{\text{up-parabola flow }(-)}\\ \underbrace{(\mathrm{nF},\mathrm{DP})}_{\text{down-parabola flow }(-)}\end{array}\right\}$$

for $\alpha_{j_1j_2} > 0, G_{j_1j_10} > 0$;

$$\left\{\begin{array}{c}\underbrace{(\mathrm{pF},\mathrm{UP})}_{\text{up-parabola flow }(+)}\\ \underbrace{(\mathrm{UP},\mathrm{pF})}_{\text{up-parabola flow }(+)}\end{array}\right\} \rightleftharpoons \underbrace{(\mathrm{DI},\mathrm{SI})}_{\text{siwthcing bifurcation}} \rightleftharpoons \left\{\begin{array}{c}\underbrace{(\mathrm{DP},\mathrm{nF})}_{\text{down-parabola flow }(-)}\\ \underbrace{(\mathrm{nF},\mathrm{DP})}_{\text{down-parabola flow }(-)}\end{array}\right\}$$

for $\alpha_{j_1j_2} < 0, G_{j_1j_10} < 0$.

- For $a_{j_1j_10}\big(a_{j_1j_1s}-a_{j_1j_1l}\big)>0$ and $a_{j_2j_20}<0$,

$$\left\{\begin{array}{c}\underbrace{(\mathrm{nF},\mathrm{UP})}_{\text{up-parabola flow }(-)}\\ \underbrace{(\mathrm{DP},\mathrm{nF})}_{\text{down-parabola flow }(-)}\end{array}\right\}\rightleftharpoons\underbrace{(\mathrm{DI},\mathrm{SO})}_{\text{switching bifurcartion}}\rightleftharpoons\left\{\begin{array}{c}\underbrace{(\mathrm{UP},\mathrm{pF})}_{\text{up-parabola flow }(+)}\\ \underbrace{(\mathrm{pF},\mathrm{DP})}_{\text{down-parabola flow }(+)}\end{array}\right\}$$

for $\alpha_{j_1j_2}>0, G_{j_1j_10}>0$;

$$\left\{\begin{array}{c}\underbrace{(\mathrm{nF},\mathrm{UP})}_{\text{up-parabola flow }(-)}\\ \underbrace{(\mathrm{DP},\mathrm{nF})}_{\text{down-parabola flow }(-)}\end{array}\right\}\rightleftharpoons\underbrace{(\mathrm{DI},\mathrm{SI})}_{\text{switching bifurcartion}}\rightleftharpoons\left\{\begin{array}{c}\underbrace{(\mathrm{UP},\mathrm{pF})}_{\text{up-parabola flow }(+)}\\ \underbrace{(\mathrm{pF},\mathrm{DP})}_{\text{down-parabola flow }(+)}\end{array}\right\}$$

for $\alpha_{j_1j_2}>0, G_{j_1j_10}<0$;

$$\left\{\begin{array}{c}\underbrace{(\mathrm{nF},\mathrm{UP})}_{\text{up-parabola flow }(-)}\\ \underbrace{(\mathrm{UP},\mathrm{nF})}_{\text{up-parabola flow }(-)}\end{array}\right\}\rightleftharpoons\underbrace{(\mathrm{DI},\mathrm{SO})}_{\text{switching bifurcation}}\rightleftharpoons\left\{\begin{array}{c}\underbrace{(\mathrm{DP},\mathrm{pF})}_{\text{down-parabola flow }(+)}\\ \underbrace{(\mathrm{pF},\mathrm{DP})}_{\text{down-parabola flow }(+)}\end{array}\right\}$$

for $\alpha_{j_1j_2}<0, G_{j_1j_10}>0$.

- For $a_{j_1j_10}(a_{j_1j_1s}-a_{j_1j_1l})<0$ and $a_{j_2j_20}<0$,

$$\left\{\begin{array}{c}\underbrace{(\mathrm{pF},\mathrm{DP})}_{\text{down-parabola flow }(+)}\\ \underbrace{(\mathrm{UP},\mathrm{nF})}_{\text{up-parabola flow }(-)}\end{array}\right\}\rightleftharpoons\underbrace{(\mathrm{SI},\mathrm{II})}_{\text{switching bifurcation}}\rightleftharpoons\left\{\begin{array}{c}\underbrace{(\mathrm{DP},\mathrm{pF})}_{\text{down-parabola flow }(+)}\\ \underbrace{(\mathrm{nF},\mathrm{UP})}_{\text{up-parabola flow }(-)}\end{array}\right\}$$

for $\alpha_{j_1j_2}>0, G_{j_1j_10}<0$;

$$\left\{\begin{array}{c}\underbrace{(\mathrm{pF},\mathrm{DP})}_{\text{down-parabola flow }(+)}\\ \underbrace{(\mathrm{DP},\mathrm{nF})}_{\text{down-parabola flow }(-)}\end{array}\right\}\rightleftharpoons\underbrace{(\mathrm{SI},\mathrm{II})}_{\text{switching bifurcation}}\rightleftharpoons\left\{\begin{array}{c}\underbrace{(\mathrm{UP},\mathrm{pF})}_{\text{up-parabola flow }(+)}\\ \underbrace{(\mathrm{nF},\mathrm{UP})}_{\text{up-parabola flow }(-)}\end{array}\right\}$$

for $\alpha_{j_1j_2}<0, G_{j_1j_10}<0$;

$$\left\{\begin{array}{c}\underbrace{(\mathrm{pF},\mathrm{DP})}_{\text{down-parabola flow }(+)}\\ \underbrace{(\mathrm{DP},\mathrm{nF})}_{\text{down-parabola flow }(-)}\end{array}\right\}\rightleftharpoons\underbrace{(\mathrm{SI},\mathrm{II})}_{\text{switching bifurcation}}\rightleftharpoons\left\{\begin{array}{c}\underbrace{(\mathrm{UP},\mathrm{pF})}_{\text{up-parabola flow }(+)}\\ \underbrace{(\mathrm{nF},\mathrm{UP})}_{\text{up-parabola flow }(-)}\end{array}\right\}$$

for $\alpha_{j_1j_2}<0, G_{j_1j_10}>0$.

(v) From cases (ii$_2$), (iii$_1$), and (iii$_2$), the equilibrium of $y_{j_1}^* = a_{j_1 j_1 1}$ *for the* x_{j_1}*-direction has four switching bifurcations* as in Eqs. (3.72)–(3.75).

- For $a_{j_1 j_1 0} > 0$ and $a_{j_2 j_2 0} > 0$,

$$\underbrace{(\mathrm{pF},\mathrm{UP})}_{\text{up-parabola flow }(+)} \rightleftharpoons \underbrace{(\mathrm{UP},\mathrm{SO})}_{\text{switching bifurcation}} \rightleftharpoons \left\{\begin{array}{c} \underbrace{(\mathrm{UP},\mathrm{pF})}_{\text{up-parabola flow }(+)} \\ \underbrace{(\mathrm{nF},\mathrm{DP})}_{\text{down-parabola flow }(-)} \\ \underbrace{(\mathrm{UP},\mathrm{nF})}_{\text{up-parabola flow }(-)} \end{array}\right\} \text{for } \alpha_{j_1 j_2} > 0,$$

$$\underbrace{(\mathrm{pF},\mathrm{UP})}_{\text{up-parabola flow }(+)} \rightleftharpoons \underbrace{(\mathrm{DP},\mathrm{SI})}_{\text{switching bifurcation}} \rightleftharpoons \left\{\begin{array}{c} \underbrace{(\mathrm{DP},\mathrm{pF})}_{\text{down-parabola flow }(+)} \\ \underbrace{(\mathrm{nF},\mathrm{DP})}_{\text{down-parabola flow }(-)} \\ \underbrace{(\mathrm{DP},\mathrm{nF})}_{\text{down-parabola flow }(-)} \end{array}\right\} \text{for } \alpha_{j_1 j_2} < 0.$$

- For $a_{j_1 j_1 0} < 0$ and $a_{j_2 j_2 0} > 0$,

$$\underbrace{(\mathrm{nF},\mathrm{DP})}_{\text{down-parabola flow }(-)} \rightleftharpoons \underbrace{(\mathrm{DP},\mathrm{SO})}_{\text{switching bifurcation}} \rightleftharpoons \left\{\begin{array}{c} \underbrace{(\mathrm{DP},\mathrm{nF})}_{\text{down-parabola flow }(-)} \\ \underbrace{(\mathrm{pF},\mathrm{UP})}_{\text{up-parabola flow }(+)} \\ \underbrace{(\mathrm{DP},\mathrm{nF})}_{\text{down-parabola flow }(-)} \end{array}\right\} \text{for } \alpha_{j_1 j_2} > 0,$$

$$\underbrace{(\mathrm{nF},\mathrm{DP})}_{\text{down-parabola flow }(-)} \rightleftharpoons \underbrace{(\mathrm{UP},\mathrm{SI})}_{\text{switching bifurcation}} \rightleftharpoons \left\{\begin{array}{c} \underbrace{(\mathrm{UP},\mathrm{pF})}_{\text{up-parabola flow }(+)} \\ \underbrace{(\mathrm{pF},\mathrm{UP})}_{\text{up-parabola flow }(+)} \\ \underbrace{(\mathrm{UP},\mathrm{nF})}_{\text{up-parabola flow }(-)} \end{array}\right\} \text{for } \alpha_{j_1 j_2} < 0.$$

- For $a_{j_1j_10} > 0$ and $a_{j_2j_20} < 0$,

$$\underbrace{(\mathrm{pF}, \mathrm{DP})}_{\text{down-parabola flow }(-)} \rightleftharpoons \underbrace{(\mathrm{DP}, \mathrm{SI})}_{\text{switching bifurcation}} \rightleftharpoons \left\{ \begin{array}{c} \underbrace{(\mathrm{DP}, \mathrm{nF})}_{\text{down-parabola flow }(-)} \\ \underbrace{(\mathrm{nF}, \mathrm{UP})}_{\text{up-parabola flow }(-)} \\ \underbrace{(\mathrm{DP}, \mathrm{pF})}_{\text{down-parabola flow }(+)} \end{array} \right\} \text{for } \alpha_{j_1j_2} > 0,$$

$$\underbrace{(\mathrm{pF}, \mathrm{DP})}_{\text{down-parabola flow }(-)} \rightleftharpoons \underbrace{(\mathrm{UP}, \mathrm{SO})}_{\text{switching bifurcation}} \rightleftharpoons \left\{ \begin{array}{c} \underbrace{(\mathrm{UP}, \mathrm{nF})}_{\text{up-parabola flow }(-)} \\ \underbrace{(\mathrm{nF}, \mathrm{UP})}_{\text{up-parabola flow }(-)} \\ \underbrace{(\mathrm{UP}, \mathrm{pF})}_{\text{up-parabola flow }(+)} \end{array} \right\} \text{for } \alpha_{j_1j_2} < 0.$$

- For $a_{j_1j_10} < 0$ and $a_{j_2j_20} < 0$,

$$\underbrace{(\mathrm{nF}, \mathrm{UP})}_{\text{up-parabola flow }(-)} \rightleftharpoons \underbrace{(\mathrm{UP}, \mathrm{SI})}_{\text{switching bifurcation}} \rightleftharpoons \left\{ \begin{array}{c} \underbrace{(\mathrm{UP}, \mathrm{nF})}_{\text{up-parabola flow }(-)} \\ \underbrace{(\mathrm{pF}, \mathrm{DP})}_{\text{down-parabola flow }(+)} \\ \underbrace{(\mathrm{UP}, \mathrm{pF})}_{\text{up-parabola flow }(+)} \end{array} \right\} \text{for } \alpha_{j_1j_2} > 0,$$

$$\underbrace{(\mathrm{nF}, \mathrm{UP})}_{\text{up-parabola flow }(-)} \rightleftharpoons \underbrace{(\mathrm{DP}, \mathrm{SO})}_{\text{switching bifurcation}} \rightleftharpoons \left\{ \begin{array}{c} \underbrace{(\mathrm{DP}, \mathrm{nF})}_{\text{down-parabola flow }(-)} \\ \underbrace{(\mathrm{pF}, \mathrm{DP})}_{\text{down-parabola flow }(+)} \\ \underbrace{(\mathrm{DP}, \mathrm{pF})}_{\text{down-parabola flow }(+)} \end{array} \right\} \text{for } \alpha_{j_1j_2} < 0.$$

In the end, this theorem is proved. ■

3.3 Parabola-Flow Switching and Appearing

In this section, the flow switching and appearing bifurcations in the two-dimensional single-bivariate quadratic dynamical systems will be discussed as in Luo (2023).

3.3.1 *Inflection-Flows for Parabola-Flows Appearing*

The inflection flows are the appearing bifurcations of up- and down-parabola flows in the same direction.

Consider a two-dimensional system as

$$\begin{aligned}\dot{x}_1 &= a_{110}(y_1 - a_{111})^2,\\ \dot{x}_2 &= a_{220}(y_1 - a_{221}),\end{aligned} \tag{3.76}$$

with

$$y_1 = \alpha_{11}x_1 + \alpha_{12}x_2. \tag{3.77}$$

There is a dynamical system in the y_1-direction as

$$\dot{y}_1 = \alpha_{11}a_{110}\left[(y_1 - a_{111})^2 + c_{11}(y_1 - a_{111}) + c_{11}(a_{111} - a_{221})\right], \tag{3.78}$$

with a dynamical system in the $y_1^{\perp}$-direction as

$$\dot{y}_1^{\perp} = \beta_{11}a_{110}(y_1 - a_{111})^2 + \beta_{12}a_{220}(y_1 - a_{221}), \tag{3.79}$$

where

$$\begin{aligned}&y_1^{\perp} = \beta_{11}x_1 + \beta_{12}x_2, \mathbf{n}_1^{\perp} = (\beta_{11}, \beta_{12})^{\mathrm{T}};\\ &(\mathbf{n}_1)^{\mathrm{T}} \cdot \mathbf{n}_1^{\perp} = \alpha_{11}\beta_{11} + \alpha_{12}\beta_{12} = 0;\\ &c_{11} = \alpha_{11}a_{110}/\alpha_{12}a_{220}.\end{aligned} \tag{3.80}$$

Consider

$$\Delta = (c_{11})^2 - 4c_{11}(a_{111} - a_{221}). \tag{3.81}$$

- For $\Delta < 0$, there is no equilibriums in the y_1-direction,
- For $\Delta = 0$, there is one equilibriums in the y_1-direction,
- For $\Delta > 0$, there are two equilibriums in the y_1-direction.

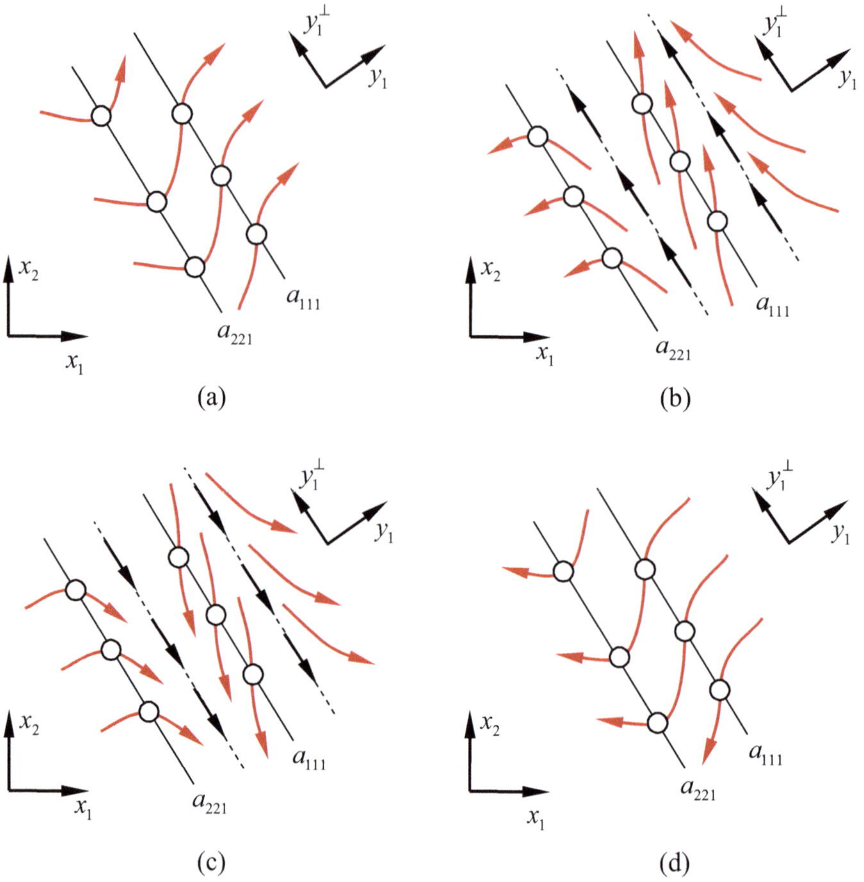

Fig. 3.1 Phase portraits ($a_{111} > a_{221}$ and $\alpha_{11} > 0$, $\alpha_{12} > 0$) for two-dimensional systems on the x_1-direction with $y_1^* = a_{111}$ and on the x_2-direction with $y_1^* = a_{221}$: (**a**) (II,pF)-flow of $y_1^* = a_{111}$ and (pF, UP)-flow of $y_1^* = a_{221}$ ($a_{110} > 0$, $a_{220} > 0$), (**b**) (DI,pF)-flow of $y_1^* = a_{111}$ and (nF,DP)-flow of $y_1^* = a_{221}$ ($a_{110} < 0$, $a_{220} > 0$), (**c**) (DI,nF)-flow of $y_1^* = a_{111}$ and (pF,DP)-flow of $y_1^* = a_{221}$ ($a_{110} > 0$, $a_{220} < 0$), (**d**) (II,nF)-flow of $y_1^* = a_{111}$ and (nF,UP)-flow of $y_1^* = a_{221}$ ($a_{110} < 0$, $a_{220} < 0$)

For $a_{111} > a_{221}$ with $\alpha_{11} > 0$ and $\alpha_{12} > 0$, presented in Fig. 3.1 are the one-dimensional flows of $(y_1^*, \dot{x}_2) = (a_{111}, \dot{x}_2)$ and $(\dot{x}_1, y_1^*) = (\dot{x}_1, a_{221})$. In Fig. 3.1a, the (II,pF)-flow of $y_1^* = a_{111}$ and (pF,UP)-flow of $y_1^* = a_{221}$ ($a_{110} > 0$, $a_{220} > 0$) are presented. Owing to $c_{11} > 0$, $\Delta > 0$ if $c_{11} > 4(a_{111} - a_{221})$. Thus there are two equilibriums in the y_1-direction. $\Delta < 0$ if $c_{11} < 4(a_{111} - a_{221})$. So no any equilibrium exists in the y_1-direction. $\Delta = 0$ if $c_{11} = 4(a_{111} - a_{221})$. Thus, there is an equilibrium. $\Delta < 0$ if $c_{11} < 4(a_{111} - a_{221})$. Herein, it is assumed no equilibrium exists. In Fig. 3.1b, the (DI,pF)-flow of $y_1^* = a_{111}$ and (nF,DP)-flow of $y_1^* = a_{221}$ ($a_{110} < 0$, $a_{220} > 0$) are presented. Owing to $c_{11} < 0$, $\Delta > 0$. Thus there are two equilibriums in the y_1-direction. The one-dimensional flows of two equilibriums in the y_1-direction are positive sink and source flows. In Fig. 3.1c, the (DI,nF)-flow of $y_1^* = a_{111}$ and

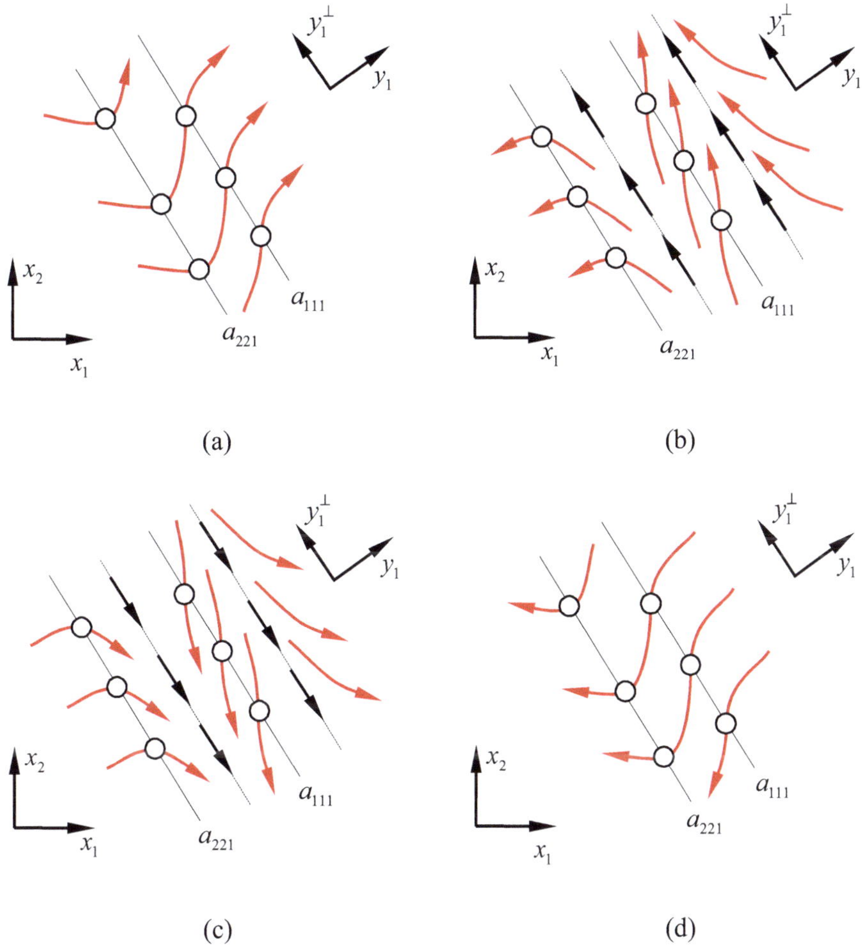

Fig. 3.2 Phase portraits ($a_{111} > a_{221}$ and $\alpha_{11} > 0$, $\alpha_{12} < 0$) for two-dimensional systems on the x_1-direction with $y_1^* = a_{111}$ and on the x_2-direction with $y_1^* = a_{221}$: **(a)** (II,pF)-flow of $y_1^* = a_{111}$ and (pF, UP)-flow of $y_1^* = a_{221}$ ($a_{110} > 0$, $a_{220} > 0$), **(b)** (DI,pF)-flow of $y_1^* = a_{111}$ and (nF,DP)-flow of $y_1^* = a_{221}$ ($a_{110} < 0$, $a_{220} > 0$), **(c)** (DI,nF)-flow of $y_1^* = a_{111}$ and (pF,DP)-flow of $y_1^* = a_{221}$ ($a_{110} > 0$, $a_{220} < 0$), **(d)** (II,nF)-flow of $y_1^* = a_{111}$ and (nF,UP)-flow of $y_1^* = a_{221}$ ($a_{110} < 0$, $a_{220} < 0$)

(pF,DP)-flow of $y_1^* = a_{221}$ ($a_{110} > 0$, $a_{220} < 0$) are presented. Because of $c_{11} < 0$, we have $\Delta > 0$. Thus there are also two equilibriums in the y_1-direction. The one-dimensional flows of two equilibriums in the y_1-direction are negative sink and source flows. In Fig. 3.1d, the (II,nF)-flow of $y_1^* = a_{111}$ and (nF,UP)-flow of $y_1^* = a_{221}$ ($a_{110} < 0$, $a_{220} < 0$) are sketched. Owing to $c_{11} > 0$, it is also assumed no equilibrium exists.

For $a_{111} > a_{221}$ with $\alpha_{11} > 0$ and $\alpha_{12} < 0$, presented in Fig. 3.2 are the one-dimensional flows of $(y_1^*, \dot{x}_2) = (a_{111}, \dot{x}_2)$ and $(\dot{x}_1, y_1^*) = (\dot{x}_1, a_{221})$. In Fig. 3.2a, the (II,pF)-flow of $y_1^* = a_{111}$ and (pF,UP)-flow of $y_1^* = a_{221}$ ($a_{110} > 0$, $a_{220} > 0$) are presented. Owing to $c_{11} < 0$, $\Delta > 0$. Thus there are two equilibriums in the y_1-direction. The one-dimensional flows of two equilibriums in the y_1-direction are positive sink and source flows. In Fig. 3.2b, the (DI,pF)-flow of $y_1^* = a_{111}$ and (nF, DP)-flow of $y_1^* = a_{221}$ ($a_{110} < 0$, $a_{220} > 0$) are presented. Owing to $c_{11} > 0$, $\Delta > 0$ if $c_{11} > 4(a_{111} - a_{221})$ and there are two equilibriums in the y_1-direction. $\Delta < 0$ if $c_{11} < 4(a_{111} - a_{221})$ and no any equilibrium exists in the y_1-direction. $\Delta = 0$ if $c_{11} = 4(a_{111} - a_{221})$ and there is an equilibrium. $\Delta < 0$ if $c_{11} < 4(a_{111} - a_{221})$. Herein, it is assumed no equilibrium exists. In Fig. 3.2c, the (DI,nF)-flow of $y_1^* = a_{111}$ and (pF,DP)-flow of $y_1^* = a_{221}$ ($a_{110} > 0$, $a_{220} < 0$) are presented. Owing to $c_{11} > 0$, it was also assumed no equilibrium exists. In Fig. 3.2d, the (II,nF)-flow of $y_1^* = a_{111}$ and (nF,UP)-flow of $y_1^* = a_{221}$ ($a_{110} < 0$, $a_{220} < 0$) are sketched. Because of $c_{11} < 0$, we have $\Delta > 0$ and two equilibriums exist in the y_1-direction. The one-dimensional flows of two equilibriums in the y_1-direction are negative sink and source flows. The appearing bifurcations are as follows.

- For $a_{110} > 0$ and $a_{220} > 0$,

$$\begin{aligned}
&\underbrace{(\text{II},\text{pF})}_{\text{appearing bifurcation}} \rightleftharpoons \left\{\begin{array}{c}\underbrace{(\text{DP},\text{pF})}_{\text{down-parabola flow }(+)}\\ \underbrace{(\text{UP},\text{pF})}_{\text{up-parabola flow }(+)}\end{array}\right\} \text{for } \alpha_{12} > 0;\\
&\underbrace{(\text{II},\text{pF})}_{\text{appearing bifurcation}} \rightleftharpoons \left\{\begin{array}{c}\underbrace{(\text{UP},\text{pF})}_{\text{up-parabola flow }(+)}\\ \underbrace{(\text{DP},\text{pF})}_{\text{down-parabola flow }(+)}\end{array}\right\} \text{for } \alpha_{12} < 0.
\end{aligned} \tag{3.82}$$

- For $a_{110} < 0$ and $a_{220} > 0$,

$$\begin{aligned}
&\underbrace{(\text{DI},\text{pF})}_{\text{appearing bifurcation}} \rightleftharpoons \left\{\begin{array}{c}\underbrace{(\text{UP},\text{pF})}_{\text{up-parabola flow }(+)}\\ \underbrace{(\text{DP},\text{pF})}_{\text{down-parabola flow }(+)}\end{array}\right\} \text{for } \alpha_{12} > 0;\\
&\underbrace{(\text{DI},\text{pF})}_{\text{appearing bifurcation}} \rightleftharpoons \left\{\begin{array}{c}\underbrace{(\text{DP},\text{pF})}_{\text{down-parabola flow }(+)}\\ \underbrace{(\text{UP},\text{pF})}_{\text{up-parabola flow }(+)}\end{array}\right\} \text{for } \alpha_{12} < 0.
\end{aligned} \tag{3.83}$$

- For $a_{110} > 0$ and $a_{220} < 0$,

$$\begin{aligned}
&\underbrace{(\mathrm{DI},\mathrm{nF})}_{\text{appearing bifurcation}} \rightleftharpoons \left\{\begin{matrix} \underbrace{(\mathrm{UP},\mathrm{nF})}_{\text{up-parabola flow }(-)} \\ \underbrace{(\mathrm{DP},\mathrm{nF})}_{\text{down-parabola flow }(-)} \end{matrix}\right\} \text{for } \alpha_{12} > 0; \\
&\underbrace{(\mathrm{DI},\mathrm{nF})}_{\text{appearing bifurcation}} \rightleftharpoons \left\{\begin{matrix} \underbrace{(\mathrm{DP},\mathrm{nF})}_{\text{down-parabola flow }(-)} \\ \underbrace{(\mathrm{UP},\mathrm{nF})}_{\text{up-parabola flow }(-)} \end{matrix}\right\} \text{for } \alpha_{12} < 0.
\end{aligned} \tag{3.84}$$

- For $a_{110} < 0$ and $a_{220} < 0$,

$$\begin{aligned}
&\underbrace{(\mathrm{II},\mathrm{nF})}_{\text{appearing bifurcation}} \rightleftharpoons \left\{\begin{matrix} \underbrace{(\mathrm{DP},\mathrm{nF})}_{\text{up-parabola flow }(-)} \\ \underbrace{(\mathrm{UP},\mathrm{nF})}_{\text{down-parabola flow }(-)} \end{matrix}\right\} \text{for } \alpha_{12} > 0; \\
&\underbrace{(\mathrm{II},\mathrm{nF})}_{\text{appearing bifurcation}} \rightleftharpoons \left\{\begin{matrix} \underbrace{(\mathrm{UP},\mathrm{nF})}_{\text{up-parabola flow }(-)} \\ \underbrace{(\mathrm{DP},\mathrm{nF})}_{\text{down-parabola flow }(-)} \end{matrix}\right\} \text{for } \alpha_{12} > 0.
\end{aligned} \tag{3.85}$$

For $a_{111} < a_{221}$ with $\alpha_{11} > 0$ and $\alpha_{12} > 0$, the one-dimensional flows of $(y_1^*, \dot{x}_2) = (a_{111}, \dot{x}_2)$ and $(\dot{x}_1, y_1^*) = (\dot{x}_1, a_{221})$ are presented in Fig. 3.3. In Fig. 3.3a, the (DI,nF)-flow of $y_1^* = a_{111}$ and (pF,UP)-flow of $y_1^* = a_{221}$ ($a_{110} > 0$, $a_{220} > 0$) are given. Owing to $c_{11} < 0$, $\Delta > 0$ and there are two equilibriums in the y_1-direction. The one-dimensional flows of two equilibriums in the y_1-direction are positive sink and source flows. In Fig. 3.3b, the (II,nF)-flow of $y_1^* = a_{111}$ and (nF,DP)-flow of $y_1^* = a_{221}$ ($a_{110} < 0$, $a_{220} > 0$) are presented. Owing to $c_{11} > 0$, it is assumed no equilibrium exists. In Fig. 3.3c, the (II,pF)-flow of $y_1^* = a_{111}$ and (pF,DP)-flow of $y_1^* = a_{221}$ ($a_{110} > 0$, $a_{220} < 0$) are presented. Owing to $c_{11} > 0$, it was also assumed no equilibrium exists. In Fig. 3.3d, the (DI,pF)-flow of $y_1^* = a_{111}$ and (nF,UP)-flow of $y_1^* = a_{221}$ ($a_{110} < 0$, $a_{220} < 0$) are sketched. Because of $c_{11} < 0$, $\Delta > 0$ and two equilibriums exist in the y_1-direction. The one-dimensional flows of two equilibriums in the y_1-direction are negative sink and source flows.

For $a_{111} < a_{221}$ with $\alpha_{11} > 0$ and $\alpha_{12} < 0$, presented in Fig. 3.4 are the one-dimensional flows of $(y_1^*, \dot{x}_2) = (a_{111}, \dot{x}_2)$ and $(\dot{x}_1, y_1^*) = (\dot{x}_1, a_{221})$. In Fig. 3.4a,

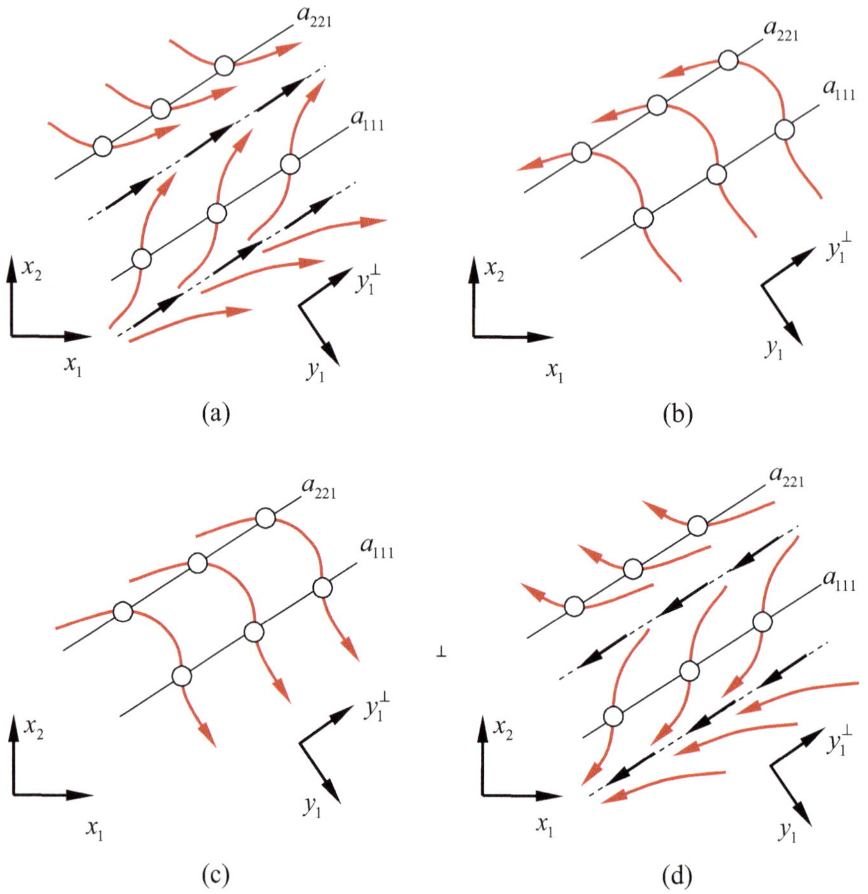

Fig. 3.3 Phase portraits ($a_{111} < a_{221}$ and $\alpha_{11} > 0$, $\alpha_{12} > 0$) for two-dimensional systems on the x_1-direction with $y_1^* = a_{111}$ and on the x_2-direction with $y_1^* = a_{221}$: (**a**) (DI,nF)-flow of $y_1^* = a_{111}$ and (pF,UP)-flow of $y_1^* = a_{221}$ ($a_{110} > 0$, $a_{220} > 0$), (**b**) (II,nF)-flow of $y_1^* = a_{111}$ and (nF,DP)-flow of $y_1^* = a_{221}$ ($a_{110} < 0$, $a_{220} > 0$), (**c**) (II,pF)-flow of $y_1^* = a_{111}$ and (pF,DP)-flow of $y_1^* = a_{221}$ ($a_{110} > 0$, $a_{220} < 0$), (**d**) (DI,pF)-flow of $y_1^* = a_{111}$ and (nF,UP)-flow of $y_1^* = a_{221}$ ($a_{110} < 0$, $a_{220} < 0$)

the (DI,nF)-flow of $y_1^* = a_{111}$ and (pF,UP)-flow of $y_1^* = a_{221}$ ($a_{110} > 0$, $a_{220} > 0$) are presented. Owing to $c_{11} > 0$, it is assumed no equilibrium exists. In Fig. 3.4b, the (II, nF)-flow of $y_1^* = a_{111}$ and (nF,DP)-flow of $y_1^* = a_{221}$ ($a_{110} < 0$, $a_{220} > 0$) are presented. Owing to $c_{11} < 0$, $\Delta > 0$ and there are two equilibriums in the y_1-direction, which are depicted in the y_1-direction as the positive sink and source flows. In Fig. 3.4c, the (II,pF)-flow of $y_1^* = a_{111}$ and (pF,DP)-flow of $y_1^* = a_{221}$ ($a_{110} > 0$,

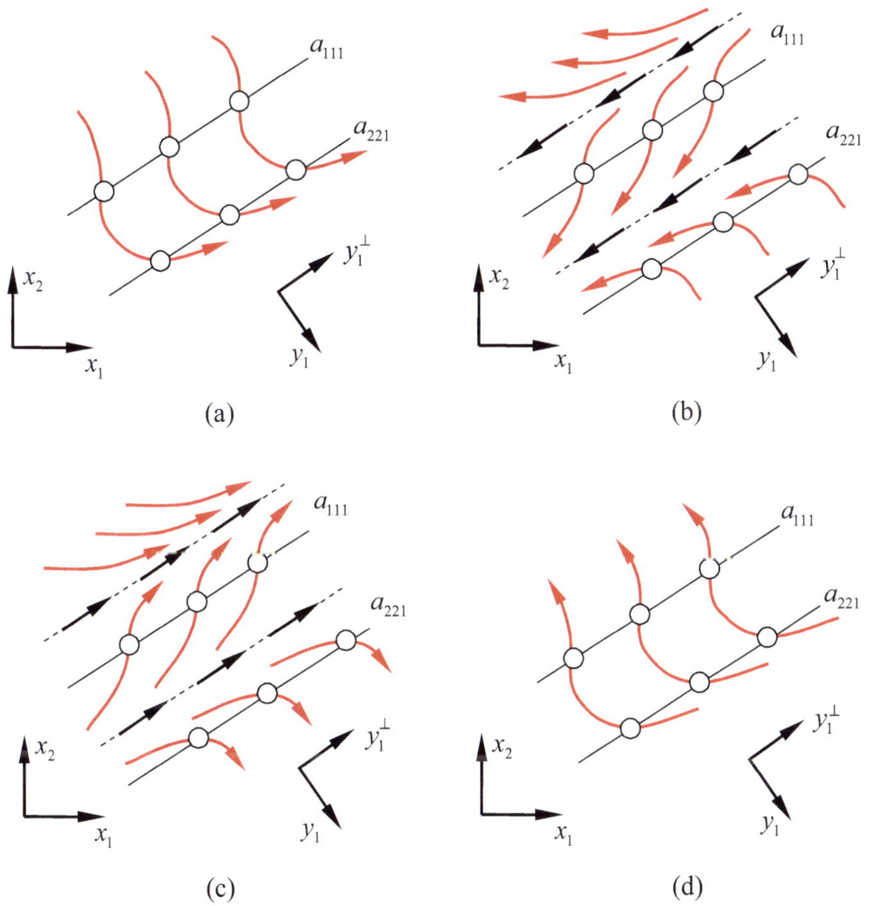

Fig. 3.4 Phase portraits ($a_{111} < a_{221}$ and $\alpha_{11} > 0$, $\alpha_{12} < 0$) for two-dimensional systems on the x_1-direction with $y_1^* = a_{111}$ and on the x_2-direction with $y_1^* = a_{221}$: (**a**) (DI,nF)-flow of $y_1^* = a_{111}$ and (pF,UP)-flow of $y_1^* = a_{221}$ ($a_{110} > 0$, $a_{220} > 0$), (**b**) (II,nF)-flow of $y_1^* = a_{111}$ and (nF,DP)-flow of $y_1^* = a_{221}$ ($a_{110} < 0$, $a_{220} > 0$), (**c**) (II,pF)-flow of $y_1^* = a_{111}$ and (pF,DP)-flow of $y_1^* = a_{221}$ ($a_{110} > 0$, $a_{220} < 0$), (**d**) (DI,pF)-flow of $y_1^* = a_{111}$ and (nF,UP)-flow of $y_1^* = a_{221}$ ($a_{110} < 0$, $a_{220} < 0$)

$a_{220} < 0$) are presented. Because of $c_{11} < 0$, $\Delta > 0$ and there are also two equilibriums in the y_1-direction, which are sketched in the y_1-direction as the negative sink and source flows. In Fig. 3.4d, the (DI,nF)-flow of $y_1^* = a_{111}$ and (nF, UP)-flow of $y_1^* = a_{221}$ ($a_{110} < 0$, $a_{220} < 0$) are sketched. Owing to $c_{11} > 0$, it is also assumed no equilibrium exists.

The corresponding appearing bifurcations are given as follows.

- For $a_{110} > 0$ and $a_{220} > 0$,

$$\begin{aligned}&\underbrace{(\mathrm{DI},\mathrm{nF})}_{\text{appearing bifurcation}} \rightleftharpoons \left\{\begin{matrix}\underbrace{(\mathrm{UP},\mathrm{nF})}_{\text{up-parabola flow }(-)}\\ \underbrace{(\mathrm{DP},\mathrm{nF})}_{\text{down-parabola flow }(-)}\end{matrix}\right\} \text{ for } \alpha_{12} > 0;\\ &\underbrace{(\mathrm{DI},\mathrm{nF})}_{\text{appearing bifurcation}} \rightleftharpoons \left\{\begin{matrix}\underbrace{(\mathrm{DP},\mathrm{nF})}_{\text{down-parabola flow }(-)}\\ \underbrace{(\mathrm{UP},\mathrm{nF})}_{\text{up-parabola flow }(-)}\end{matrix}\right\} \text{ for } \alpha_{12} < 0.\end{aligned} \tag{3.86}$$

- For $a_{110} < 0$ and $a_{220} > 0$,

$$\begin{aligned}&\underbrace{(\mathrm{II},\mathrm{nF})}_{\text{appearing bifurcation}} \rightleftharpoons \left\{\begin{matrix}\underbrace{(\mathrm{DP},\mathrm{nF})}_{\text{down-parabola flow }(-)}\\ \underbrace{(\mathrm{UP},\mathrm{nF})}_{\text{up-parabola flow }(-)}\end{matrix}\right\} \text{ for } \alpha_{12} > 0;\\ &\underbrace{(\mathrm{II},\mathrm{nF})}_{\text{appearing bifurcation}} \rightleftharpoons \left\{\begin{matrix}\underbrace{(\mathrm{UP},\mathrm{nF})}_{\text{up-parabola flow }(-)}\\ \underbrace{(\mathrm{DP},\mathrm{nF})}_{\text{down-parabola flow }(-)}\end{matrix}\right\} \text{ for } \alpha_{12} < 0.\end{aligned} \tag{3.87}$$

- For $a_{110} > 0$ and $a_{220} < 0$,

$$\begin{aligned}&\underbrace{(\mathrm{II},\mathrm{pF})}_{\text{appearing bifurcation}} \rightleftharpoons \left\{\begin{matrix}\underbrace{(\mathrm{DP},\mathrm{pF})}_{\text{down-parabola flow }(+)}\\ \underbrace{(\mathrm{UP},\mathrm{pF})}_{\text{up-parabola flow }(+)}\end{matrix}\right\} \text{ for } \alpha_{12} > 0;\\ &\underbrace{(\mathrm{II},\mathrm{pF})}_{\text{appearing bifurcation}} \rightleftharpoons \left\{\begin{matrix}\underbrace{(\mathrm{UP},\mathrm{pF})}_{\text{up-parabola flow }(+)}\\ \underbrace{(\mathrm{DP},\mathrm{pF})}_{\text{down-parabola flow }(+)}\end{matrix}\right\} \text{ for } \alpha_{12} < 0.\end{aligned} \tag{3.88}$$

- For $a_{110} < 0$ and $a_{220} < 0$,

$$\underbrace{(\text{DI, pF})}_{\text{appearing bifurcation}} \rightleftharpoons \left\{ \begin{array}{c} \underbrace{(\text{UP, pF})}_{\text{up-parabola flow } (+)} \\ \underbrace{(\text{DP, pF})}_{\text{down-parabola flow } (+)} \end{array} \right\} \text{for } \alpha_{12} > 0;$$
$$\underbrace{(\text{DI, pF})}_{\text{appearing bifurcation}} \rightleftharpoons \left\{ \begin{array}{c} \underbrace{(\text{DP, pF})}_{\text{down-parabola flow } (+)} \\ \underbrace{(\text{UP, pF})}_{\text{up-parabola flow } (+)} \end{array} \right\} \text{for } \alpha_{12} < 0. \tag{3.89}$$

3.3.2 Parabola-Source and Sink Infinite-Equilibriums

The parabola infinite-equilibriums are the switching bifurcations between inflection and parabola flows.

Consider a two-dimensional system for $a_{111} = a_{221}$ as

$$\begin{aligned} \dot{x}_1 &= a_{110}(y_1 - a_{111})^2, \\ \dot{x}_2 &= a_{220}(y_1 - a_{221}). \end{aligned} \tag{3.90}$$

There is a dynamical system in the y_1-direction as

$$\dot{y}_1 = \alpha_{11} a_{110}(y_1 - a_{111})(y_1 - a_{111} + c_{11}), \tag{3.91}$$

with a dynamical system in the $y_1^\perp$-direction as

$$\dot{y}_1^\perp = \beta_{11} a_{110}(y_1 - a_{111})(y_1 - a_{111} + c_{11}^\perp), \tag{3.92}$$

where

$$c_{11} = \alpha_{11} a_{110} / \alpha_{12} a_{220}, c_{11}^\perp = \beta_{11} a_{110} / \beta_{12} a_{220}. \tag{3.93}$$

The infinite-equilibrium of $y_1^* = a_{111}$ is a switching bifurcation as follows.

- For $a_{j_1j_10} > 0$ and $a_{j_2j_20} > 0$,

$$\left\{\begin{array}{c}\underbrace{(\mathrm{pF},\mathrm{UP})}_{\text{up-parabola flow }(+)}\\ \underbrace{(\mathrm{II},\mathrm{pF})}_{\text{increasing-inflection flow }(+)}\end{array}\right\} \rightleftharpoons \underbrace{(\mathrm{UP},\mathrm{SO})}_{\text{switching bifurcation}} \rightleftharpoons \left\{\begin{array}{c}\underbrace{(\mathrm{DI},\mathrm{nF})}_{\text{decreasing-inflection flow }(-)}\\ \underbrace{(\mathrm{pF},\mathrm{UP})}_{\text{up-parabola flow }(+)}\end{array}\right\}. \tag{3.94}$$

- For $a_{j_1j_10} < 0$ and $a_{j_2j_20} > 0$,

$$\left\{\begin{array}{c}\underbrace{(\mathrm{nF},\mathrm{DP})}_{\text{down-parabola flow }(-)}\\ \underbrace{(\mathrm{DI},\mathrm{pF})}_{\text{decreasing-inflection flow }(+)}\end{array}\right\} \rightleftharpoons \underbrace{(\mathrm{DP},\mathrm{SO})}_{\text{switching bifurcation}} \rightleftharpoons \left\{\begin{array}{c}\underbrace{(\mathrm{II},\mathrm{nF})}_{\text{increasing-inflection flow }(-)}\\ \underbrace{(\mathrm{nF},\mathrm{DP})}_{\text{down-parabola flow }(-)}\end{array}\right\}. \tag{3.95}$$

- For $a_{110} > 0$ and $a_{220} < 0$,

$$\left\{\begin{array}{c}\underbrace{(\mathrm{pF},\mathrm{DP})}_{\text{down-parabola flow }(+)}\\ \underbrace{(\mathrm{DI},\mathrm{nF})}_{\text{decreasing-inflection flow }(-)}\end{array}\right\} \rightleftharpoons \underbrace{(\mathrm{DP},\mathrm{SI})}_{\text{switching bifurcation}} \rightleftharpoons \left\{\begin{array}{c}\underbrace{(\mathrm{II},\mathrm{pF})}_{\text{increasing-inflection flow }(+)}\\ \underbrace{(\mathrm{pF},\mathrm{DP})}_{\text{down-parabola flow }(+)}\end{array}\right\}. \tag{3.96}$$

- For $a_{110} < 0$ and $a_{220} < 0$,

$$\left\{\begin{array}{c}\underbrace{(\mathrm{nF},\mathrm{UP})}_{\text{up-parabola flow }(-)}\\ \underbrace{(\mathrm{II},\mathrm{nF})}_{\text{increasing-inflection flow }(-)}\end{array}\right\} \rightleftharpoons \underbrace{(\mathrm{UP},\mathrm{SI})}_{\text{switching bifurcation}} \rightleftharpoons \left\{\begin{array}{c}\underbrace{(\mathrm{DI},\mathrm{pF})}_{\text{decreasing-inflection flow }(+)}\\ \underbrace{(\mathrm{nF},\mathrm{UP})}_{\text{up-parabola flow }(-)}\end{array}\right\}. \tag{3.97}$$

The infinite-equilibriums for ($\alpha_{11} > 0$, $\alpha_{12} > 0$) are presented in Fig. 3.5a–d for the (UP,SO)-infinite equilibrium ($a_{110} > 0$, $a_{220} > 0$), (DP,SO)-infinite equilibrium ($a_{110} < 0$, $a_{220} > 0$), (DP,SI)-infinite equilibrium ($a_{110} < 0$, $a_{220} > 0$), and (UP,SI)-infinite equilibrium ($a_{110} < 0$, $a_{220} < 0$), respectively. The corresponding equilibriums in the y_1-direction are presented as well. For ($a_{110} > 0$, $a_{220} > 0$), the negative source flow of $y_1 = a_{111} - c_{11}$ is depicted. For ($a_{110} < 0$, $a_{220} > 0$), the positive source flow of $y_1 = a_{111} - c_{11}$ is shown. For ($a_{110} > 0$, $a_{220} < 0$), the positive sink flow of $y_1 = a_{111} - c_{11}$ exists. For ($a_{110} < 0$, $a_{220} < 0$), the positive sink flow of $y_1 = a_{111} - c_{11}$ exists. The infinite-equilibriums for ($\alpha_{11} > 0$, $\alpha_{12} < 0$) are presented in Fig. 3.6. The types of infinite-equilibriums are the same as for ($\alpha_{11} > 0$, $\alpha_{12} > 0$). Only sink and source flows of $y_1 = a_{111} - c_{11}$ are different. For ($a_{110} > 0$, $a_{220} > 0$), the positive sink flow exists. For ($a_{110} < 0$, $a_{220} > 0$), the negative sink flow exists. For ($a_{110} > 0$, $a_{220} < 0$), the positive source flow exists. For ($a_{110} < 0$, $a_{220} < 0$), the negative source flow exists.

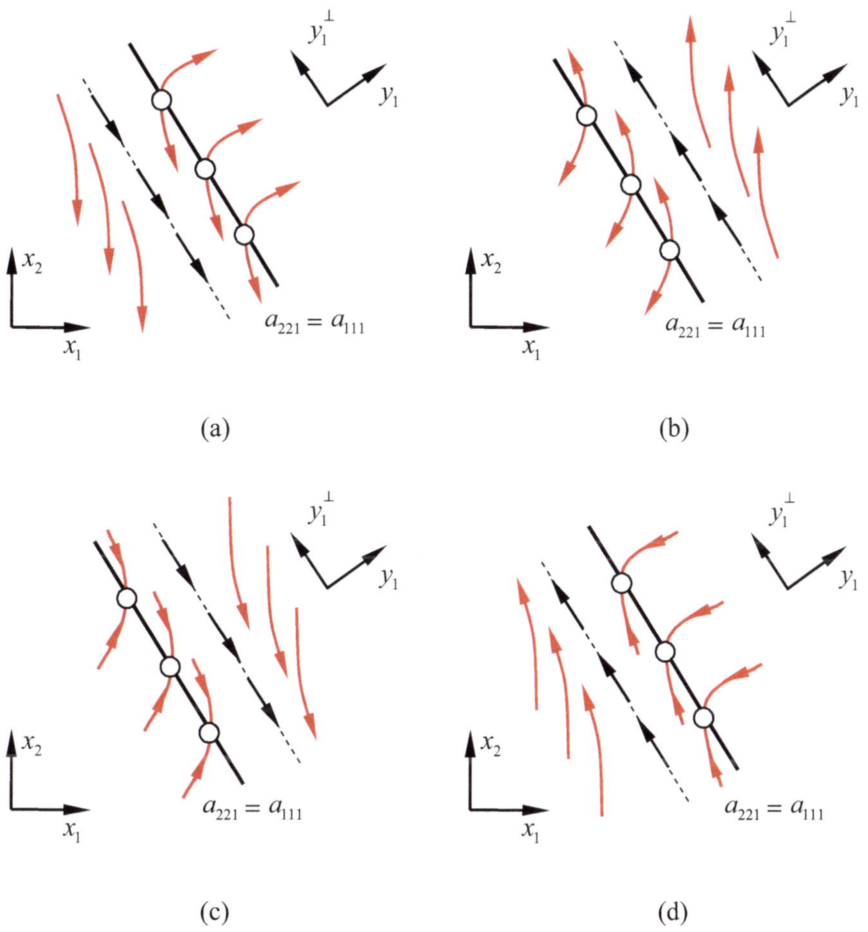

Fig. 3.5 Phase portraits ($\alpha_{11} > 0$, $\alpha_{12} > 0$) for two-dimensional systems on the x_1-direction with $y_1^* = a_{111}$ and on the x_2-direction with $y_1^* = a_{111}$: (**a**) (UP,SO)-infinite equilibrium ($a_{110} > 0$, $a_{220} > 0$), (**b**) (DP,SO)-infinite equilibrium ($a_{110} < 0$, $a_{220} > 0$), (**c**) (DP,SI)-infinite equilibrium ($a_{110} < 0$, $a_{220} > 0$), (**d**) (UP,SI)-infinite equilibrium ($a_{110} < 0$, $a_{220} < 0$)

3.3.3 Simple Parabola-Flows Arrays

Consider a dynamical system as

$$\begin{aligned} \dot{x}_2 &= a_{220}(y_1 - a_{221}), \\ \dot{x}_1 &= a_{111}(y_1 - a_{111})(y_1 - a_{112}). \end{aligned} \tag{3.98}$$

There is a dynamical system in the y_1-direction as

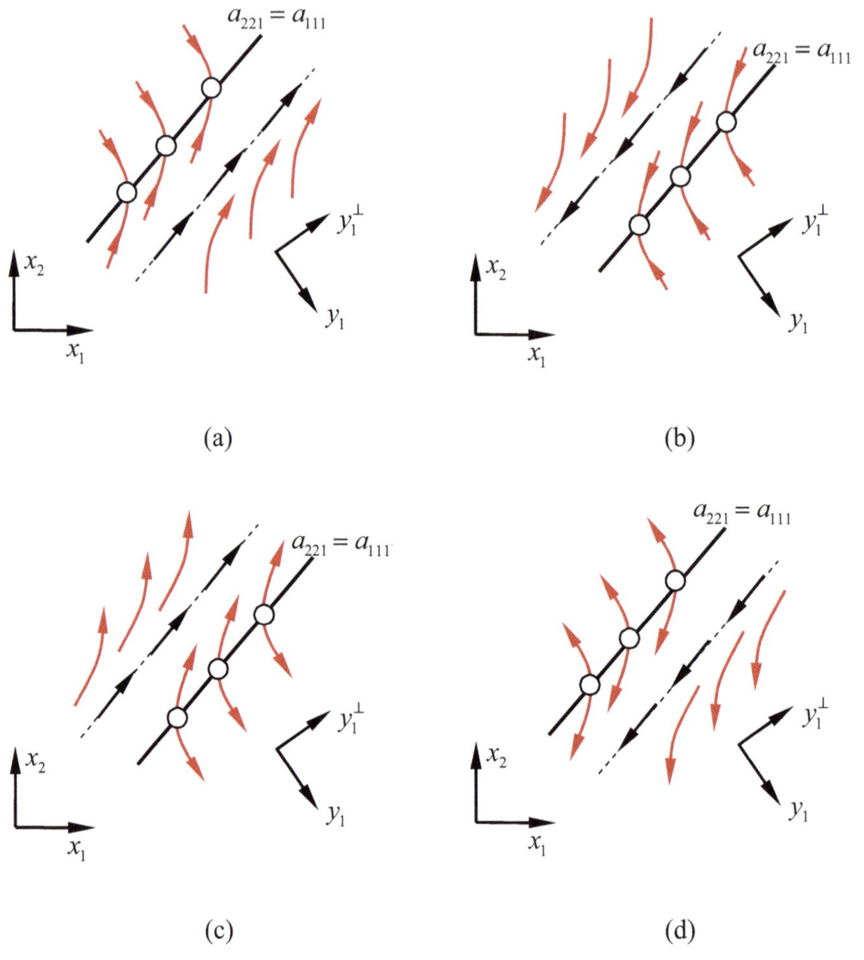

Fig. 3.6 Phase portraits ($\alpha_{11} > 0$, $\alpha_{12} < 0$)for two-dimensional systems on the x_1-direction with $y_1^* = a_{111}$ and on the x_2-direction with $y_1^* = a_{111}$: (**a**) (DP,SI)-infinite-equilibrium ($a_{110} > 0$, $a_{220} > 0$), (**b**) (UP,SI)-infinite-equilibrium ($a_{110} < 0$, $a_{220} > 0$), (**c**) (UP,SO)-infinite equilibrium ($a_{110} < 0$, $a_{220} > 0$), (**d**) (DP,SO)-infinite-equilibrium ($a_{110} < 0$, $a_{220} < 0$)

$$\dot{y}_1 = \alpha_{11}a_{110}\left[y_1^2 - (a_{111} + a_{112} - c_{11})y_1 + (a_{111}a_{112} - c_{11}a_{221})\right] \tag{3.99}$$

with a dynamical system in the $y_1^{\perp}$-direction as

$$\dot{y}_1^{\perp} = \beta_{11}a_{110}(y_1 - a_{111})(y_1 - a_{112}) + \beta_{12}a_{220}(y_1 - a_{221}), \tag{3.100}$$

and

$$\begin{aligned}\Delta &= (a_{111} + a_{112} - c_{11})^2 - 4(a_{111}a_{112} - c_{11}a_{221}) \\ &= (a_{111} - a_{112} - c_{11})^2 + 4c_{11}(a_{211} - a_{112}) \\ &= (a_{112} - a_{111} - c_{11})^2 + 4c_{11}(a_{221} - a_{111}).\end{aligned} \quad (3.101)$$

- For $\Delta < 0$, there is no equilibrium in the y_1-direction.
- For $\Delta = 0$, there is an equilibrium in the y_1-direction for the one-dimensional lower-saddle or upper-saddle flows.
- For $\Delta > 0$, there are two equilibriums in the y_1-direction for the one-dimensional sink and source flows.

It is assumed that the sink and source flows in the y_1-direction exist. The phase portraits for two-dimensional systems on the x_1-direction with $y_1^* = a_{111}, a_{112}$ and on the x_2-direction with $y_1^* = a_{221}$ are presented in Figs. 3.7–3.10. The dash inclined lines are for the sink and source flows in the y_1-direction. The solid inclined lines are for equilibriums on the x_1- and x_2- directions.

For ($a_{111} > a_{221}$ and $\alpha_{11} > 0$, $\alpha_{12} > 0$), phase portraits for two-dimensional systems on the x_1-direction with $y_1^* = a_{111}, a_{112}$ and on the x_2-direction with $y_1^* = a_{221}$ are presented in Fig. 3.7a–d with the (DP,pF) and (UP,pF)-flows of $y_1^* = a_{111}, a_{112}$ and (pF,UP)-flow of $y_1^* = a_{221}$ ($a_{110} > 0$, $a_{220} > 0$), the (UP,pF) and (DP,pF)-flows of $y_1^* = a_{111}, a_{112}$ and (nF,DP)-flow of $y_1^* = a_{221}$ ($a_{110} < 0$, $a_{220} > 0$), the (UP,nF) and (DP,nF)-flows of $y_1^* = a_{111}, a_{112}$ and (pF,DP)-flow of $y_1^* = a_{221}$ ($a_{110} > 0$, $a_{220} < 0$), and the (DP,nF) and (UP,pF)-flows of $y_1^* = a_{111}, a_{112}$ and (nF,UP)-flow of $y_1^* = a_{221}$ ($a_{110} < 0$, $a_{220} < 0$), respectively.

For ($a_{111} > a_{221}$ and $\alpha_{11} > 0$, $\alpha_{12} < 0$), the phase portraits for two-dimensional systems on the x_1-direction with $y_1^* = a_{111}, a_{112}$ and on the x_2-direction with $y_1^* = a_{221}$ are sketched in Fig. 3.8a–d, for the (UP,pF) and (DP,pF)-flows of $y_1^* = a_{111}, a_{112}$ and (pF,UP)-flow of $y_1^* = a_{221}$ ($a_{110} > 0$, $a_{220} > 0$), the (DP,pF) and (UP,pF)-flows of $y_1^* = a_{111}, a_{112}$ and (nF,DP)-flow of $y_1^* = a_{221}$ ($a_{110} < 0$, $a_{220} > 0$), the (DP,nF) and (UP,nF)-flows of $y_1^* = a_{111}, a_{112}$ and (pF,DP)-flow of $y_1^* = a_{221}$ ($a_{110} > 0$, $a_{220} < 0$), and the (UP,nF) and (DP,nF)-flows of $y_1^* = a_{111}, a_{112}$ and (nF,UP)-flow of $y_1^* = a_{221}$ ($a_{110} < 0$, $a_{220} < 0$), respectively.

For ($a_{221} > a_{111}$ and $\alpha_{11} > 0$, $\alpha_{12} > 0$), the phase portraits for two-dimensional systems on the x_1-direction with $y_1^* = a_{111}, a_{112}$ and on the x_2-direction with $y_1^* = a_{221}$ are shown in Fig. 3.9a–d for the (UP,nF) and (DP,nF)-flows of $y_1^* = a_{111}, a_{112}$ and (pF,UP)-flow of $y_1^* = a_{221}$ ($a_{110} > 0$, $a_{220} > 0$), the (DP,pF) and (UP,pF)-flows of $y_1^* = a_{111}, a_{112}$ and (pF,DP)-flow of $y_1^* = a_{221}$ ($a_{110} < 0$, $a_{220} > 0$), the (DP,nF) and (UP,nF)-flows of $y_1^* = a_{111}, a_{112}$ and (nF,DP)-flow of $y_1^* = a_{221}$ ($a_{110} > 0$, $a_{220} < 0$), and the (UP,pF) and (DP,pF)-flows of $y_1^* = a_{111}, a_{112}$ and (nF,UP)-flow of $y_1^* = a_{221}$ ($a_{110} < 0$, $a_{220} < 0$), respectively.

For ($a_{221} > a_{112}$ and $\alpha_{11} > 0$, $\alpha_{12} < 0$), the phase portraits for two-dimensional systems on the x_1-direction with $y_1^* = a_{111}, a_{112}$ and on the x_2-direction with $y_1^* = a_{221}$ are presented in Fig. 3.10a–d with the (DP,nF) and (UP,nF)-flows of $y_1^* = a_{111}, a_{112}$ and (pF,UP)-flow of $y_1^* = a_{221}$ ($a_{110} > 0$, $a_{220} > 0$), the (UP,nF) and (DP,nF)-flows of $y_1^* = a_{111}, a_{112}$ and (nF,DP)-flow of $y_1^* = a_{221}$ ($a_{110} < 0$, $a_{220} > 0$), the (UP,pF) and

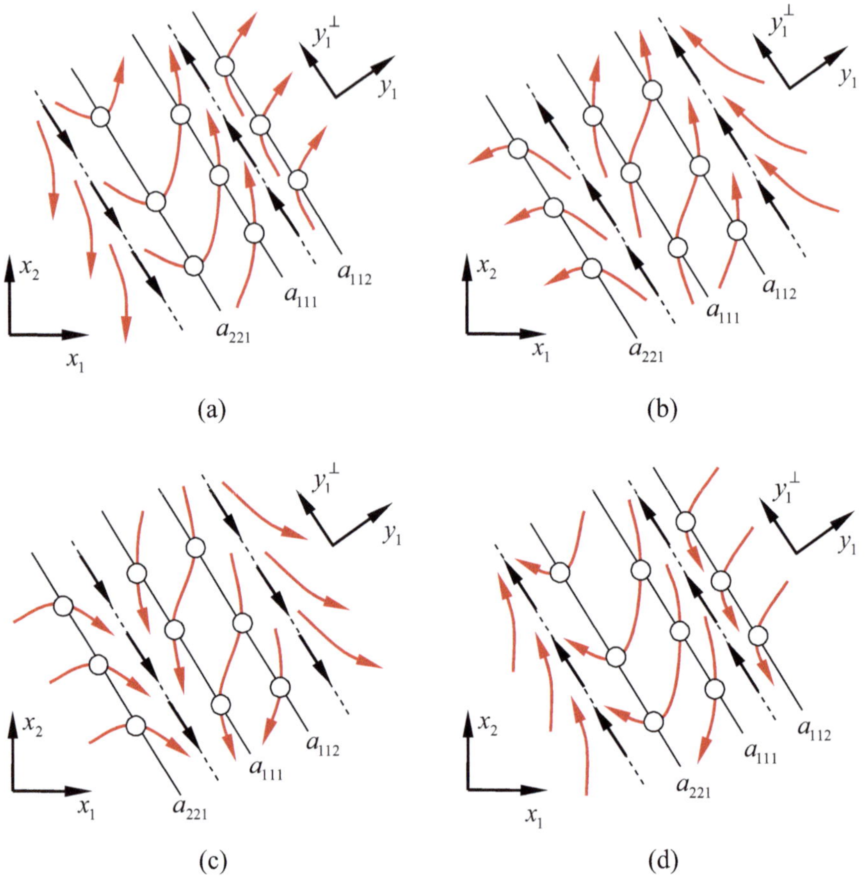

Fig. 3.7 Phase portraits ($a_{111} > a_{221}$ and $\alpha_{11} > 0$, $\alpha_{12} > 0$) for two-dimensional systems on the x_1-direction with $y_1^* = a_{111}, a_{112}$ and on the x_2-direction with $y_1^* = a_{221}$: (**a**) (DP,pF) and (UP,pF)-flows of $y_1^* = a_{111}, a_{112}$ and (pF,UP)-flow of $y_1^* = a_{221}$ ($a_{110} > 0$, $a_{220} > 0$), (**b**) (UP,pF) and (DP,pF)-flows of $y_1^* = a_{111}, a_{112}$ and (nF,DP)-flow of $y_1^* = a_{221}$ ($a_{110} < 0$, $a_{220} > 0$), (**c**) (UP,nF) and (DP,nF)-flows of $y_1^* = a_{111}, a_{112}$ and (pF,DP)-flow of $y_1^* = a_{221}$ ($a_{110} > 0$, $a_{220} < 0$), (**d**) (DP,nF) and (UP,pF)-flows of $y_1^* = a_{111}, a_{112}$ and (nF,UP)-flow of $y_1^* = a_{221}$ ($a_{110} < 0$, $a_{220} < 0$)

(DP,pF)-flows of $y_1^* = a_{111}, a_{112}$ and (pF,DP)-flow of $y_1^* = a_{221}$ ($a_{110} > 0$, $a_{220} < 0$), the (DP,pF) and (UP,pF)-flows of $y_1^* = a_{111}, a_{112}$ and (nF,UP)-flow of $y_1^* = a_{221}$ ($a_{110} < 0$, $a_{220} < 0$), respectively.

For ($a_{111} < a_{221} < a_{112}$ and $\alpha_{11} > 0$, $\alpha_{12} > 0$), the phase portraits for two-dimensional systems on the x_1-direction with $y_1^* = a_{111}, a_{112}$ and on the x_2-direction with $y_1^* = a_{221}$ are presented in Fig. 3.11a–d for the (UP,nF) and (UP, pF)-flows of $y_1^* = a_{111}, a_{112}$ and (nF,DP)-flow of $y_1^* = a_{221}$ ($a_{110} > 0$, $a_{220} > 0$), the (DP,nF) and (DP,pF)-flows of $y_1^* = a_{111}, a_{112}$ and (pF,UP)-flow of $y_1^* = a_{221}$ ($a_{110} < 0$, $a_{220} > 0$), the (DP,pF) and (DP,nF)-flows of $y_1^* = a_{111}, a_{112}$ and (nF,

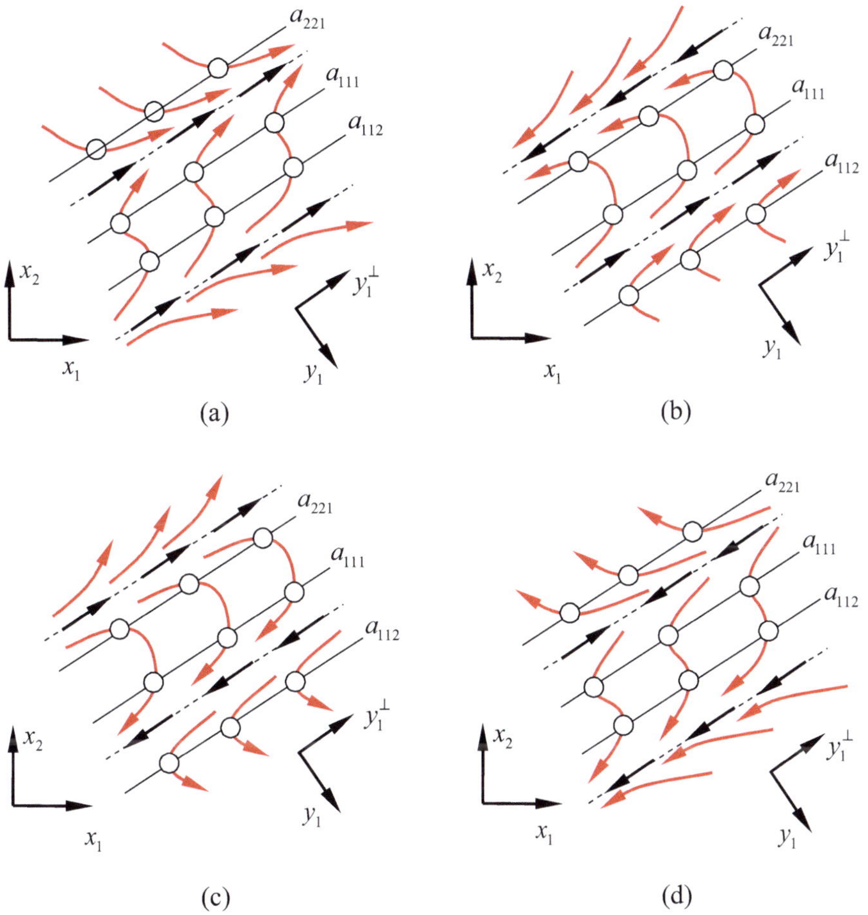

Fig. 3.8 Phase portraits ($a_{111} > a_{221}$ and $\alpha_{11} > 0$, $\alpha_{12} < 0$) for two-dimensional systems on the x_1-direction with $y_1^* = a_{111}, a_{112}$ and on the x_2-direction with $y_1^* = a_{221}$: (**a**) (UP,pF) and (DP,pF)-flows of $y_1^* = a_{111}, a_{112}$ and (pF,UP)-flow of $y_1^* = a_{221}$ ($a_{110} > 0$, $a_{220} > 0$), (**b**) (DP,pF) and (UP,pF)-flows of $y_1^* = a_{111}, a_{112}$ and (nF,DP)-flow of $y_1^* = a_{221}$ ($a_{110} < 0$, $a_{220} > 0$), (**c**) (DP,nF) and (UP,nF)-flows of $y_1^* = a_{111}, a_{112}$ and (pF,DP)-flow of $y_1^* = a_{221}$ ($a_{110} > 0$, $a_{220} < 0$), (**d**) (UP,nF) and (DP,nF)-flows of $y_1^* = a_{111}, a_{112}$ and (nF,UP)-flow of $y_1^* = a_{221}$ ($a_{110} < 0$, $a_{220} < 0$)

UP)-flow of $y_1^* = a_{221}$ ($a_{110} > 0$, $a_{220} < 0$), and the (UP,pF) and (UP,nF)-flows of $y_1^* = a_{111}, a_{112}$ and (pF,DP)-flow of $y_1^* = a_{221}$ ($a_{110} < 0$, $a_{220} < 0$), respectively.

For ($a_{111} < a_{221} < a_{112}$ and $\alpha_{11} > 0$, $\alpha_{12} < 0$), the phase portraits for two-dimensional systems on the x_1-direction with $y_1^* = a_{111}, a_{112}$ and on the x_2-direction with $y_1^* = a_{221}$ are sketched in Fig. 3.12a–d for the (DP,nF) and (DP,pF)-flows of $y_1^* = a_{111}, a_{112}$ and (nF,DP)-flow of $y_1^* = a_{221}$ ($a_{110} > 0$, $a_{220} > 0$), the (UP, nF) and (UP,pF)-flows of $y_1^* = a_{111}, a_{112}$ and (pF,UP)-flow of $y_1^* = a_{221}$ ($a_{110} < 0$, $a_{220} > 0$), the (UP,pF) and (UP,nF)-flows of $y_1^* = a_{111}, a_{112}$ and (nF,UP)-flow of

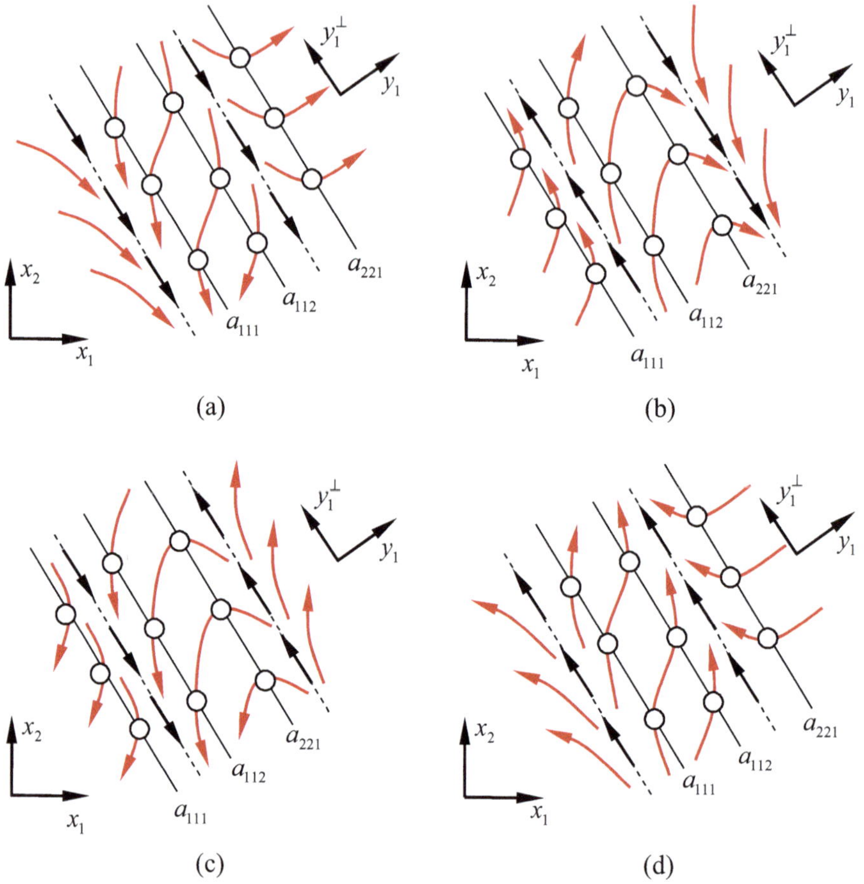

Fig. 3.9 Phase portraits ($a_{221} > a_{112}$ and $\alpha_{11} > 0$, $\alpha_{12} > 0$) for two-dimensional systems on the x_1-direction with $y_1^* = a_{111}, a_{112}$ and on the x_2-direction with $y_1^* = a_{221}$: (**a**) (UP,nF) and (DP,nF)-flows of $y_1^* = a_{111}, a_{112}$ and (pF,UP)-flow of $y_1^* = a_{221}$ ($a_{110} > 0$, $a_{220} > 0$), (**b**) (DP,pF) and (UP,pF)-flows of $y_1^* = a_{111}, a_{112}$ and (pF,DP)-flow of $y_1^* = a_{221}$ ($a_{110} < 0$, $a_{220} > 0$), (**c**) (DP,nF) and (UP,nF)-flows of $y_1^* = a_{111}, a_{112}$ and (nF,DP)-flow of $y_1^* = a_{221}$ ($a_{110} > 0$, $a_{220} < 0$), (**d**) (UP,pF) and (DP,pF)-flows of $y_1^* = a_{111}, a_{112}$ and (nF,UP)-flow of $y_1^* = a_{221}$ ($a_{110} < 0$, $a_{220} < 0$)

$y_1^* = a_{221}$ ($a_{110} > 0$, $a_{220} < 0$), and the (DP,pF) and (DP,nF)-flows of $y_1^* = a_{111}, a_{112}$ and (pF,DP)-flow of $y_1^* = a_{221}$ ($a_{110} < 0$, $a_{220} < 0$), respectively.

3.3.4 *Inflection-Sources and Sinks*

The inflection infinite-equilibriums are the switching bifurcations of parabola flows in two directions.

Consider a dynamical system as

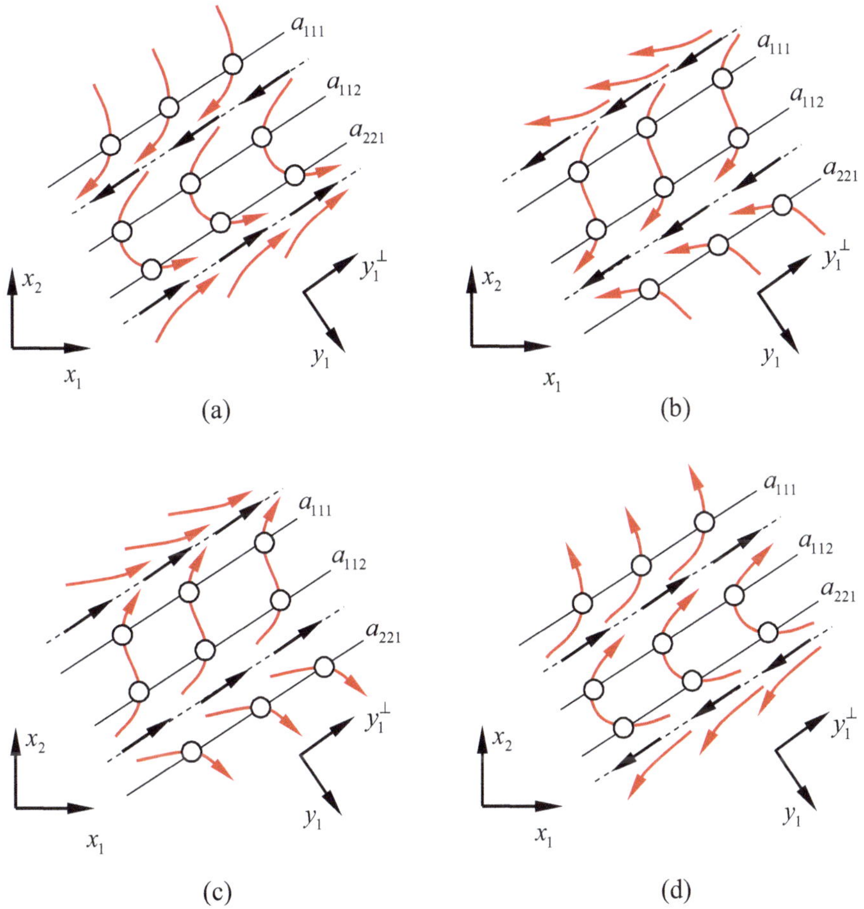

Fig. 3.10 Phase portraits ($a_{221} > a_{112}$ and $\alpha_{11} > 0$, $\alpha_{12} < 0$) for two-dimensional systems on the x_1-direction with $y_1^* = a_{111}, a_{112}$ and on the x_2-direction with $y_1^* = a_{221}$: (**a**) (DP,nF) and (UP,nF)-flows of $y_1^* = a_{111}, a_{112}$ and (pF,UP)-flow of $y_1^* = a_{221}$ ($a_{110} > 0$, $a_{220} > 0$), (**b**) (UP,nF) and (DP,nF)-flows of $y_1^* = a_{111}, a_{112}$ and (nF,DP)-flow of $y_1^* = a_{221}$ ($a_{110} < 0$, $a_{220} > 0$), (**c**) (UP,pF) and (DP,pF)-flows of $y_1^* = a_{111}, a_{112}$ and (pF,DP)-flow of $y_1^* = a_{221}$ ($a_{110} > 0$, $a_{220} < 0$), (**d**) (DP,pF) and (UP,pF)-flows of $y_1^* = a_{111}, a_{112}$ and (nF,UP)-flow of $y_1^* = a_{221}$ ($a_{110} < 0$, $a_{220} < 0$)

$$
\begin{aligned}
&\dot{x}_2 = a_{220}(y_1 - a_{221}),\\
&\dot{x}_1 = a_{111}(y_1 - a_{11s})(y_1 - a_{11l}),\\
&s, l \in \{1, 2\}, s \neq l.
\end{aligned}
\tag{3.102}
$$

For $a_{221} = a_{11s}(s = 1, 2)$, there is a dynamical system in the y_{j_1}-direction as

$$
\dot{y}_1 = \alpha_{11} a_{110}(y_1 - a_{11s})(y_1 - a_{11l} + c_{11}),
\tag{3.103}
$$

with a dynamical system in the $y_1^{\perp}$-direction as

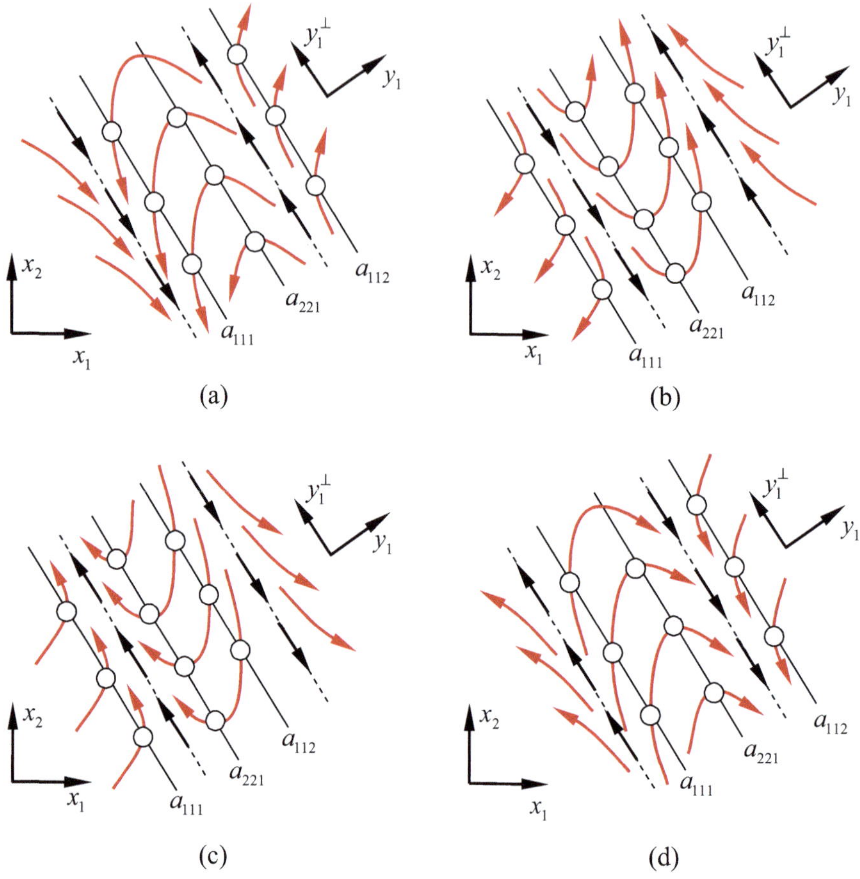

Fig. 3.11 Phase portraits ($a_{111} < a_{221} < a_{112}$ and $\alpha_{11} > 0, \alpha_{12} > 0$) for two-dimensional systems on the x_1-direction with $y_1^* = a_{111}, a_{112}$ and on the x_2-direction with $y_1^* = a_{221}$: (**a**) (UP,nF) and (UP,pF)-flows of $y_1^* = a_{111}, a_{112}$ and (nF,DP)-flow of $y_1^* = a_{221}$ ($a_{110} > 0$, $a_{220} > 0$), (**b**) (DP,nF) and (DP, pF)-flows of $y_1^* = a_{111}, a_{112}$ and (pF,UP)-flow of $y_1^* = a_{221}$ ($a_{110} < 0, a_{220} < 0$), (**c**) (DP,pF) and (DP, nF)-flows of $y_1^* = a_{111}, a_{112}$ and (nF,UP)-flow of $y_1^* = a_{221}$ ($a_{110} > 0$, $a_{220} < 0$), (**d**) (UP,pF) and (UP,nF)-flows of $y_1^* = a_{111}, a_{112}$ and (pF,DP)-flow of $y_1^* = a_{221}$ ($a_{110} < 0$, $a_{220} < 0$)

$$\dot{y}_1^\perp = (y_1 - a_{11s})[\beta_{11} a_{110}(y_1 - a_{11l}) + \beta_{12} a_{220}]. \tag{3.104}$$

The equilibrium of $y_1^* = a_{11l} - c_{11}$ on the y_{j_1}-direction is a one-dimensional sink or source flow. The one-dimensional sink and source flows are determined by $\alpha_{11} a_{110}(a_{11l} - a_{11s} - c_{11}) < 0$ and $\alpha_{11} a_{110}(a_{11l} - a_{11s} - c_{11}) > 0$, respectively.

The infinite-equilibrium of $y_1^* = a_{11s}$ ($s, l \in \{1, 2\}$ and $s \neq l$) is presented with $\dot{x}_2^* = a_{220}(a_{11l} - a_{221})$. *For* $a_{111} = a_{221}$, *with* $G_{110} = \alpha_{11} a_{110}(a_{111} - a_{112}) + \alpha_{12} a_{220}$, *we have:*

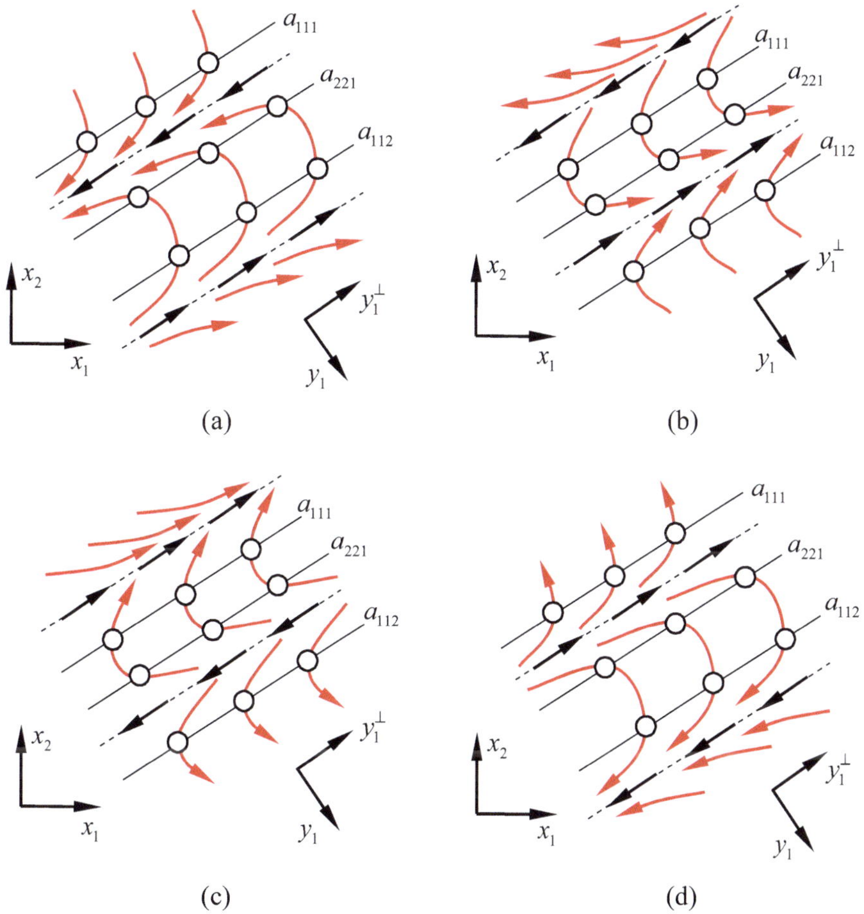

Fig. 3.12 Phase portraits ($a_{111} < a_{221} < a_{112}$ and $\alpha_{11} > 0$, $\alpha_{12} < 0$) for two-dimensional systems on the x_1-direction with $y_1^* = a_{111}, a_{112}$ and on the x_2-direction with $y_1^* = a_{221}$: (**a**) (DP,nF) and (DP,pF)-flows of $y_1^* = a_{111}, a_{112}$ and (nF,DP)-flow of $y_1^* = a_{221}$ ($a_{110} > 0$, $a_{220} > 0$), (**b**) (UP,nF) and (UP, pF)-flows of $y_1^* = a_{111}, a_{112}$ and (pF,UP)-flow of $y_1^* = a_{221}$ ($a_{110} < 0$, $a_{220} > 0$), (**c**) (UP,pF) and (UP, nF)-flows of $y_1^* = a_{111}, a_{112}$ and (nF,UP)-flow of $y_1^* = a_{221}$ ($a_{110} > 0$, $a_{220} < 0$), (**d**) (DP,pF) and (DP,nF)-flows of $y_1^* = a_{111}, a_{112}$ and (pF,DP)-flow of $y_1^* = a_{221}$ ($a_{110} < 0$, $a_{220} < 0$)

- For $a_{110} > 0$ and $a_{220} > 0$,

$$\left\{ \begin{array}{l} (a_{111}, a_{221}) \\ (a_{112}, \dot{x}_2^*) \end{array} \right\} = \left\{ \begin{array}{c} \underbrace{(\text{DI}, \text{SI})}_{\text{infinite-equilibrium}} \\ \underbrace{(\text{UP}, \text{pF})}_{\text{up-parabola flow }(+)} \end{array} \right\} \text{ for } \alpha_{12} > 0 \text{ and } G_{110} < 0,$$

$$\left\{ \begin{array}{l} (a_{111}, a_{221}) \\ (a_{112}, \dot{x}_2^*) \end{array} \right\} = \left\{ \begin{array}{c} \underbrace{(\text{DI}, \text{SO})}_{\text{infinite-equilibrium}} \\ \underbrace{(\text{UP}, \text{pF})}_{\text{up-parabola flow }(+)} \end{array} \right\} \text{ for } \alpha_{12} > 0 \text{ and } G_{110} > 0; \tag{3.105}$$

$$\left\{ \begin{array}{l} (a_{111}, a_{221}) \\ (a_{112}, \dot{x}_2^*) \end{array} \right\} = \left\{ \begin{array}{c} \underbrace{(\text{DI}, \text{SI})}_{\text{infinite-equilibrium}} \\ \underbrace{(\text{DP}, \text{pF})}_{\text{down-parabola flow }(+)} \end{array} \right\} \text{ for } \alpha_{12} < 0 \text{ and } G_{110} < 0.$$

- For $a_{110} < 0$ and $a_{220} > 0$,

$$\left\{ \begin{array}{l} (a_{111}, a_{221}) \\ (a_{112}, \dot{x}_2^*) \end{array} \right\} = \left\{ \begin{array}{c} \underbrace{(\text{II}, \text{SO})}_{\text{infinite-equilibrium}} \\ \underbrace{(\text{DP}, \text{pF})}_{\text{down-parabola flow }(+)} \end{array} \right\} \text{ for } \alpha_{12} > 0 \text{ and } G_{110} > 0;$$

$$\left\{ \begin{array}{l} (a_{111}, a_{221}) \\ (a_{112}, \dot{x}_2^*) \end{array} \right\} = \left\{ \begin{array}{c} \underbrace{(\text{II}, \text{SO})}_{\text{infinite-equilibrium}} \\ \underbrace{(\text{UP}, \text{pF})}_{\text{up-parabola flow }(+)} \end{array} \right\} \text{ for } \alpha_{12} < 0 \text{ and } G_{110} > 0, \tag{3.106}$$

$$\left\{ \begin{array}{l} (a_{111}, a_{221}) \\ (a_{112}, \dot{x}_2^*) \end{array} \right\} = \left\{ \begin{array}{c} \underbrace{(\text{II}, \text{SI})}_{\text{infinite-equilibrium}} \\ \underbrace{(\text{UP}, \text{pF})}_{\text{up-parabola flow }(+)} \end{array} \right\} \text{ for } \alpha_{12} < 0 \text{ and } G_{110} < 0.$$

- For $a_{110} > 0$ and $a_{220} < 0$,

$$\left\{\begin{matrix}(a_{111}, a_{221})\\(a_{112}, \dot{x}_2^*)\end{matrix}\right\} = \left\{\begin{matrix}\underbrace{(\mathrm{II}, \mathrm{SI})}_{\text{infinite-equilibrium}}\\\underbrace{(\mathrm{DP}, \mathrm{nF})}_{\text{down-parabola flow }(-)}\end{matrix}\right\} \text{ for } \alpha_{12} > 0 \text{ and } G_{110} < 0;$$

$$\left\{\begin{matrix}(a_{111}, a_{221})\\(a_{112}, \dot{x}_2^*)\end{matrix}\right\} = \left\{\begin{matrix}\underbrace{(\mathrm{II}, \mathrm{SI})}_{\text{infinite-equilibrium}}\\\underbrace{(\mathrm{UP}, \mathrm{nF})}_{\text{up-parabola flow }(-)}\end{matrix}\right\} \text{ for } \alpha_{12} < 0 \text{ and } G_{110} < 0, \tag{3.107}$$

$$\left\{\begin{matrix}(a_{111}, a_{221})\\(a_{112}, \dot{x}_2^*)\end{matrix}\right\} = \left\{\begin{matrix}\underbrace{(\mathrm{II}, \mathrm{SO})}_{\text{infinite-equilibrium}}\\\underbrace{(\mathrm{UP}, \mathrm{nF})}_{\text{up-parabola flow }(-)}\end{matrix}\right\} \text{ for } \alpha_{12} < 0 \text{ and } G_{110} > 0.$$

- For $a_{110} < 0$ and $a_{220} < 0$,

$$\left\{\begin{matrix}(a_{111}, a_{221})\\(a_{112}, \dot{x}_2^*)\end{matrix}\right\} = \left\{\begin{matrix}\underbrace{(\mathrm{DI}, \mathrm{SO})}_{\text{infinite-equilibrium}}\\\underbrace{(\mathrm{UP}, \mathrm{nF})}_{\text{up-parabola flow }(-)}\end{matrix}\right\} \text{ for } \alpha_{12} > 0 \text{ and } G_{110} > 0,$$

$$\left\{\begin{matrix}(a_{111}, a_{221})\\(a_{112}, \dot{x}_2^*)\end{matrix}\right\} = \left\{\begin{matrix}\underbrace{(\mathrm{DI}, \mathrm{SI})}_{\text{infinite-equilibrium}}\\\underbrace{(\mathrm{UP}, \mathrm{nF})}_{\text{up-parabola flow }(-)}\end{matrix}\right\} \text{ for } \alpha_{12} > 0 \text{ and } G_{110} < 0; \tag{3.108}$$

$$\left\{\begin{matrix}(a_{111}, a_{221})\\(a_{112}, \dot{x}_2^*)\end{matrix}\right\} = \left\{\begin{matrix}\underbrace{(\mathrm{DI}, \mathrm{SO})}_{\text{infinite-equilibrium}}\\\underbrace{(\mathrm{DP}, \mathrm{nF})}_{\text{down-parabola flow }(-)}\end{matrix}\right\} \text{ for } \alpha_{12} < 0 \text{ and } G_{110} > 0.$$

For $a_{112} = a_{221}$, *with* $G_{110} = \alpha_{11}a_{110}(a_{112} - a_{111}) + \alpha_{12}a_{220}$, we have the following cases.

- For $a_{110} > 0$ and $a_{220} > 0$,

$$\left\{\begin{array}{l}(a_{111}, \dot{x}_2^*)\\(a_{112}, a_{221})\end{array}\right\} = \left\{\begin{array}{c}\underbrace{(\text{UP, nF})}_{\text{up-parabola flow }(-)}\\\underbrace{(\text{II, SO})}_{\text{infinite-equilibrium}}\end{array}\right\} \text{for } \alpha_{12} > 0 \text{ and } G_{110} > 0;$$

$$\left\{\begin{array}{l}(a_{111}, \dot{x}_2^*)\\(a_{112}, a_{221})\end{array}\right\} = \left\{\begin{array}{c}\underbrace{(\text{DP, nF})}_{\text{down-parabola flow }(-)}\\\underbrace{(\text{II, SO})}_{\text{infinite-equilibrium}}\end{array}\right\} \text{for } \alpha_{12} < 0 \text{ and } G_{110} > 0, \quad (3.109)$$

$$\left\{\begin{array}{l}(a_{111}, \dot{x}_2^*)\\(a_{112}, a_{221})\end{array}\right\} = \left\{\begin{array}{c}\underbrace{(\text{DP, nF})}_{\text{down-parabola flow }(-)}\\\underbrace{(\text{II, SI})}_{\text{infinite-equilibrium}}\end{array}\right\} \text{for } \alpha_{12} < 0 \text{ and } G_{110} < 0.$$

- For $a_{110} < 0$ and $a_{220} > 0$,

$$\left\{\begin{array}{l}(a_{111}, \dot{x}_2^*)\\(a_{112}, a_{221})\end{array}\right\} = \left\{\begin{array}{c}\underbrace{(\text{DP, nF})}_{\text{down-parabola flow}(-)}\\\underbrace{(\text{DI, SI})}_{\text{infinite-equilibrium}}\end{array}\right\} \text{for } \alpha_{12} > 0 \text{ and } G_{110} < 0,$$

$$\left\{\begin{array}{l}(a_{111}, \dot{x}_2^*)\\(a_{112}, a_{221})\end{array}\right\} = \left\{\begin{array}{c}\underbrace{(\text{DP, nF})}_{\text{down-parabola flow }(-)}\\\underbrace{(\text{DI, SO})}_{\text{infinite-equilibrium}}\end{array}\right\} \text{for } \alpha_{12} > 0 \text{ and } G_{110} > 0; \quad (3.110)$$

$$\left\{\begin{array}{l}(a_{111}, \dot{x}_2^*)\\(a_{112}, a_{221})\end{array}\right\} = \left\{\begin{array}{c}\underbrace{(\text{UP, nF})}_{\text{up-parabola flow }(-)}\\\underbrace{(\text{SI, DI})}_{\text{infinite-equilibrium}}\end{array}\right\} \text{for } \alpha_{12} < 0 \text{ and } G_{110} < 0.$$

- For $a_{110} > 0$ and $a_{220} < 0$,

$$\begin{Bmatrix} (a_{111}, \dot{x}_2^*) \\ (a_{112}, a_{221}) \end{Bmatrix} = \begin{Bmatrix} \underbrace{(\mathrm{DP}, \mathrm{pF})}_{\text{down-parabola flow } (+)} \\ \underbrace{(\mathrm{DI}, \mathrm{SO})}_{\text{infinite-equilibrium}} \end{Bmatrix} \text{ for } \alpha_{12} > 0 \text{ and } G_{110} > 0,$$

$$\begin{Bmatrix} (a_{111}, \dot{x}_2^*) \\ (a_{112}, a_{221}) \end{Bmatrix} = \begin{Bmatrix} \underbrace{(\mathrm{DP}, \mathrm{pF})}_{\text{down-parabola flow } (+)} \\ \underbrace{(\mathrm{DI}, \mathrm{SI})}_{\text{infinite-equilibrium}} \end{Bmatrix} \text{ for } \alpha_{12} > 0 \text{ and } G_{110} < 0; \tag{3.111}$$

$$\begin{Bmatrix} (a_{111}, \dot{x}_2^*) \\ (a_{112}, a_{221}) \end{Bmatrix} = \begin{Bmatrix} \underbrace{(\mathrm{UP}, \mathrm{pF})}_{\text{up-parabola flow } (+)} \\ \underbrace{(\mathrm{DI}, \mathrm{SO})}_{\text{infinite-equilibrium}} \end{Bmatrix} \text{ for } \alpha_{12} < 0 \text{ and } G_{110} > 0.$$

- For $a_{110} < 0$ and $a_{220} < 0$,

$$\begin{Bmatrix} (a_{111}, \dot{x}_2^*) \\ (a_{112}, a_{221}) \end{Bmatrix} = \begin{Bmatrix} \underbrace{(\mathrm{UP}, \mathrm{pF})}_{\text{up-parabola flow } (+)} \\ \underbrace{(\mathrm{II}, \mathrm{SI})}_{\text{infinite-equilibrium}} \end{Bmatrix} \text{ for } \alpha_{12} > 0 \text{ and } G_{110} < 0;$$

$$\begin{Bmatrix} (a_{111}, \dot{x}_2^*) \\ (a_{112}, a_{221}) \end{Bmatrix} = \begin{Bmatrix} \underbrace{(\mathrm{DP}, \mathrm{pF})}_{\text{down-parabola flow } (+)} \\ \underbrace{(\mathrm{II}, \mathrm{SI})}_{\text{infinite-equilibrium}} \end{Bmatrix} \text{ for } \alpha_{12} < 0 \text{ and } G_{110} < 0, \tag{3.112}$$

$$\begin{Bmatrix} (a_{111}, \dot{x}_2^*) \\ (a_{112}, a_{221}) \end{Bmatrix} = \begin{Bmatrix} \underbrace{(\mathrm{DP}, \mathrm{pF})}_{\text{down-parabola flow } (+)} \\ \underbrace{(\mathrm{II}, \mathrm{SO})}_{\text{infinite-equilibrium}} \end{Bmatrix} \text{ for } \alpha_{12} < 0 \text{ and } G_{110} > 0.$$

For ($a_{111} = a_{221}$ and $\alpha_{11} > 0$, $\alpha_{12} > 0$), the phase portraits of inflection flows bifurcations for two-dimensional systems on the x_1-direction with $y_1^* = a_{111}$ and on the x_2-direction with $y_1^* = a_{221}$ is presented in Fig. 3.13a–d for the (DI,SI)-infinite-equilibrium with $G_{110} < 0$ and (DI,SO)-infinite-equilibrium with $G_{110} > 0$ for $y_1^* = a_{111} = a_{221}$ and (UP,pF)-flow of $y_1^* = a_{112}$ ($a_{110} > 0$, $a_{220} > 0$); the (II, SO)-infinite-equilibrium with $G_{110} > 0$ for $y_1^* = a_{111} = a_{221}$ and (DP,pF)-flow of $y_1^* = a_{112}$ ($a_{110} < 0$, $a_{220} > 0$); the (II,SI)-infinite-equilibrium with $G_{110} < 0$ for $y_1^* = a_{111} = a_{221}$ and (DP,nF)-flow of $y_1^* = a_{112}$ ($a_{110} > 0$, $a_{220} < 0$); the (DI,SI)-infinite-equilibrium $G_{110} < 0$ and the (DI,SO)-infinite-equilibrium $G_{110} > 0$ for $y_1^* = a_{111} = a_{221}$ and (UP,nF)-flow of $y_1^* = a_{112}$ ($a_{110} < 0$, $a_{220} < 0$), respectively.

For ($a_{111} = a_{221}$ and $\alpha_{11} > 0$, $\alpha_{12} < 0$), the phase portraits of inflection-source and sink infinite-equilibriums for two-dimensional systems on the x_1-direction with $y_1^* = a_{111}, a_{112}$ and on the x_2-direction with $y_1^* = a_{221}$ are presented in Fig. 3.14a–d for the (DI,SI)-infinite-equilibrium with $G_{110} < 0$ of $y_1^* = a_{111} = a_{221}$ and (DP,pF)-flow of $y_1^* = a_{112}$ ($a_{110} > 0$, $a_{220} > 0$); the (II,SO)-infinite-equilibrium with $G_{110} > 0$ and the (II,SI)-infinite-equilibrium with $G_{110} < 0$ for $y_1^* = a_{111} = a_{221}$ and (UP,pF)-flow of $y_1^* = a_{112}$ ($a_{110} < 0$, $a_{220} > 0$); the (II,SO)-infinite-equilibrium with $G_{110} > 0$ and the (II,SI)-infinite-equilibrium with $G_{110} < 0$ for $y_1^* = a_{111} = a_{221}$ and (UP, nF)-flow of $y_1^* = a_{112}$ ($a_{110} > 0$, $a_{220} < 0$), the (DI,SO)-infinite-equilibrium with $G_{110} > 0$ of $y_1^* = a_{111} = a_{221}$ and (DP,nF)-flow of $y_1^* = a_{112}$ ($a_{110} < 0$, $a_{220} < 0$), respectively.

For ($a_{111} = a_{221}$ and $\alpha_{11} > 0$, $\alpha_{12} > 0$), the phase portraits of inflection-source and sink infinite-equilibriums for two-dimensional systems on the x_1-direction with $y_1^* = a_{111}, a_{112}$ and on the x_2-direction with $y_1^* = a_{221}$ are presented in Fig. 3.15a–d for the (UP, nF)-flow of $y_1^* = a_{111}$ and (II,SO)-infinite-equilibrium with $G_{110} > 0$ for $y_1^* = a_{112} = a_{221}$ ($a_{110} > 0$, $a_{220} > 0$); the (DP,nF)-flow of $y_1^* = a_{111}$ and (DI,SI)-infinite-equilibrium with $G_{110} < 0$ and (DI,SO)-infinite-equilibrium with $G_{110} > 0$ of $y_1^* = a_{112} = a_{221}$ ($a_{110} < 0$, $a_{220} > 0$), the (DP,pF)-flow of $y_1^* = a_{111}$ and (DI, SI)-infinite-equilibrium with $G_{110} < 0$ and (DI,SO)-infinite-equilibrium with $G_{110} > 0$ of $y_1^* = a_{112} = a_{221}$ ($a_{110} > 0$, $a_{220} < 0$), the (UP,pF)-flow of $y_1^* = a_{111}$ and (II,SI)-infinite-equilibrium with $G_{110} < 0$ for $y_1^* = a_{112} = a_{221}$ ($a_{110} < 0$, $a_{220} < 0$).

For ($a_{112} = a_{221}$ and $\alpha_{11} > 0$, $\alpha_{12} < 0$), the phase portraits of inflection-source and sink infinite-equilibriums for two-dimensional systems on the x_1-direction with $y_1^* = a_{111}, a_{112}$ and on the x_2-direction with $y_1^* = a_{221}$ are presented in Fig. 3.16a–d for the (DP, nF)-flow of $y_1^* = a_{111}$ and (II,SO)-infinite-equilibrium with $G_{110} > 0$ and (II,SI)-infinite-equilibrium with $G_{110} < 0$ of $y_1^* = a_{112} = a_{221}$ ($a_{110} > 0$, $a_{220} > 0$); the (UP,nF)-flow of $y_1^* = a_{111}$ and (DI,SI)-infinite-equilibrium with $G_{110} < 0$ of $y_1^* = a_{112} = a_{221}$ ($a_{110} < 0$, $a_{220} > 0$); the (UP,pF)-flow of $y_1^* = a_{111}$ and (DI,SO)-infinite-equilibrium with $G_{110} > 0$ of $y_1^* = a_{112} = a_{221}$ ($a_{110} > 0$, $a_{220} < 0$), (II,SO)-infinite-equilibrium with $G_{110} > 0$ the (DP,pF)-flow of $y_1^* = a_{111}$ and (II,SI)-infinite-equilibrium with $G_{110} < 0$ and for $y_1^* = a_{112} = a_{221}$ ($a_{110} < 0$, $a_{220} < 0$).

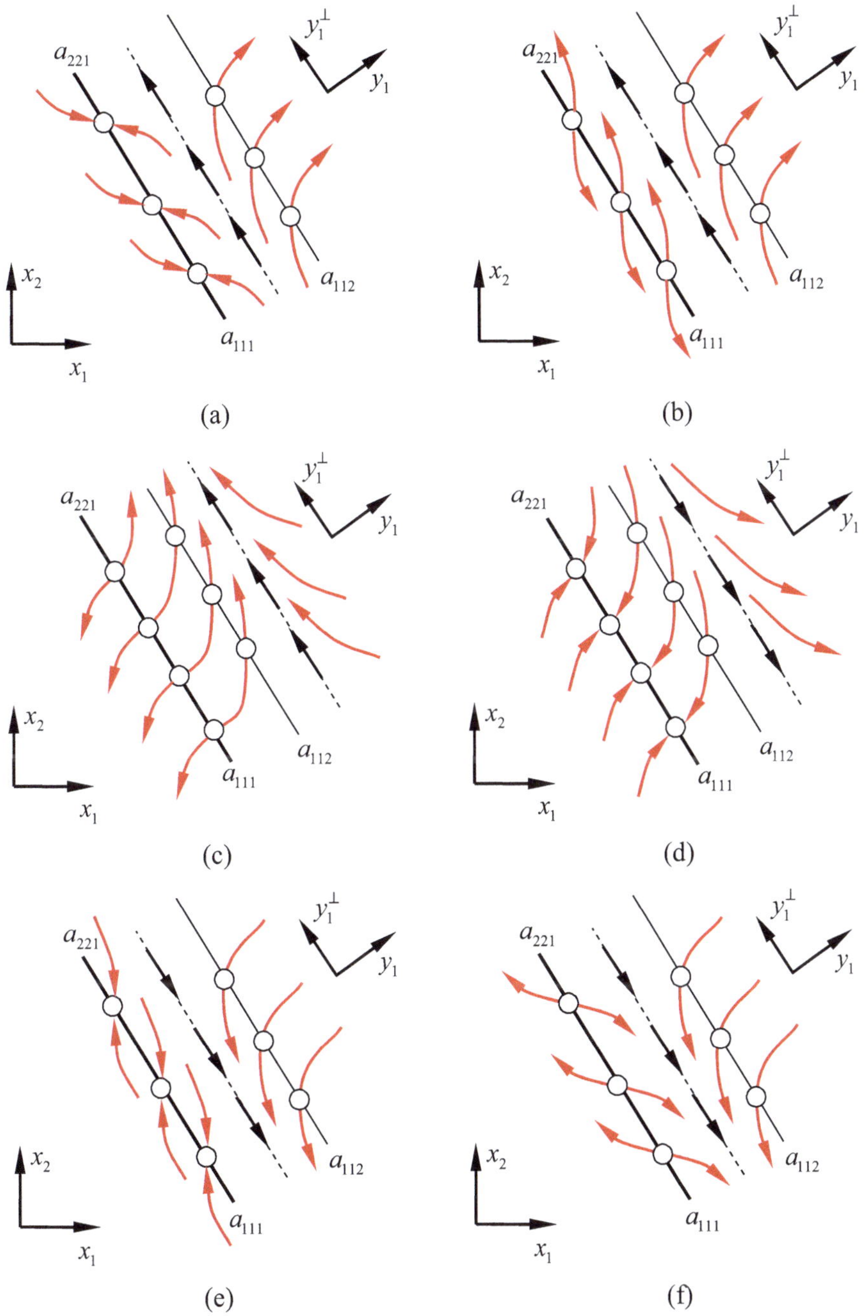

Fig. 3.13 Phase portraits ($a_{111} = a_{221}$ and $\alpha_{11} > 0$, $\alpha_{12} > 0$) for two-dimensional systems on the x_1-direction with $y_1^* = a_{111}, a_{112}$ and on the x_2-direction with $y_1^* = a_{221}$: (**a**, **b**) ($a_{110} > 0$, $a_{220} > 0$), (**c**) ($a_{110} < 0$, $a_{220} > 0$), (**d**) ($a_{110} > 0$, $a_{220} < 0$), (**e**, **f**) (($a_{110} < 0$, $a_{220} < 0$)

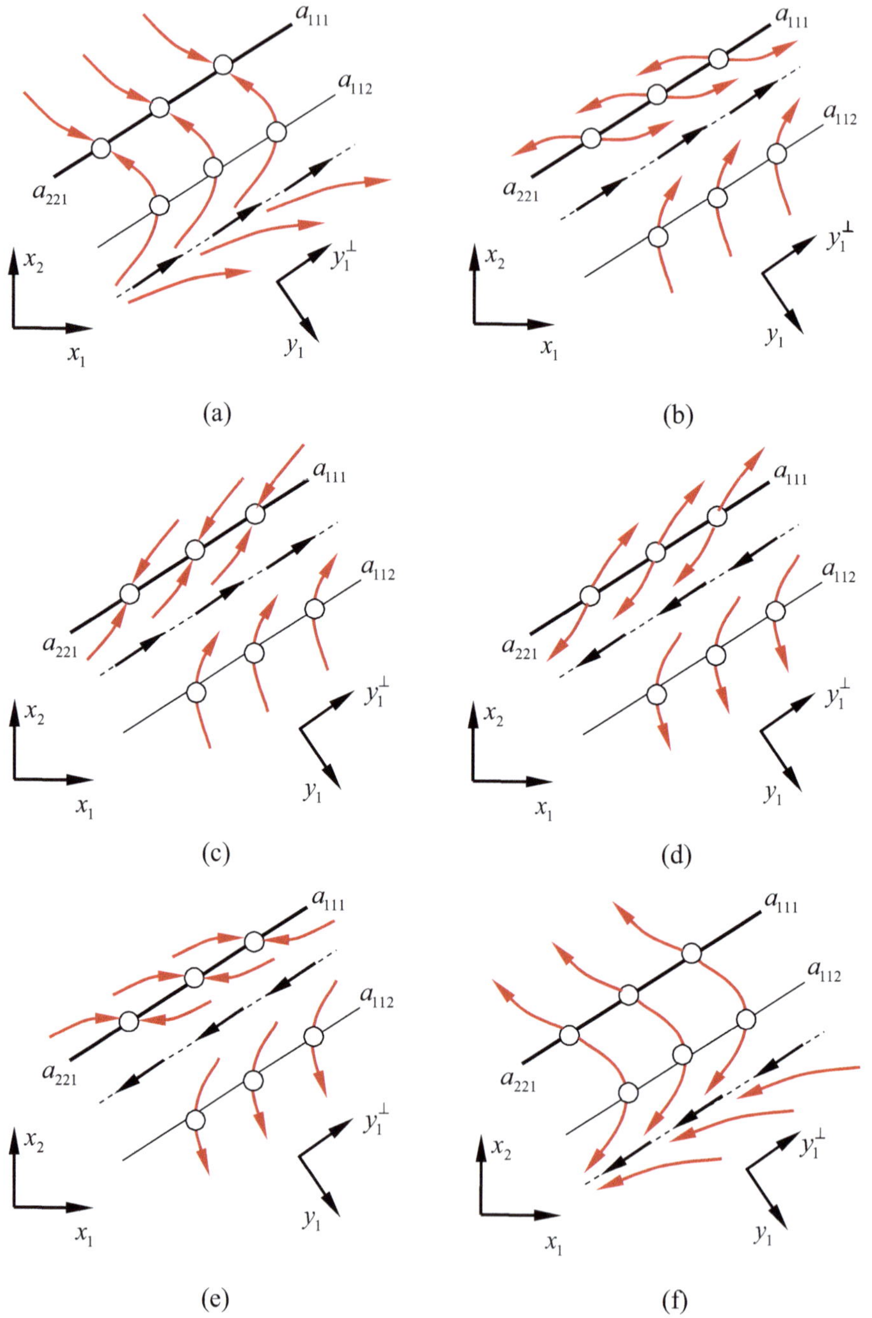

Fig. 3.14 Phase portraits ($a_{111} = a_{221}$ and $\alpha_{11} > 0$, $\alpha_{12} < 0$) for two-dimensional systems on the x_1-direction with $y_1^* = a_{111}, a_{112}$ and on the x_2-direction with $y_1^* = a_{221}$: (**a**) ($a_{110} > 0$, $a_{220} > 0$), (**b, c**) ($a_{110} < 0$, $a_{220} > 0$), (**d, e**) ($a_{110} > 0$, $a_{220} < 0$), (**d**) ($a_{110} < 0$, $a_{220} < 0$)

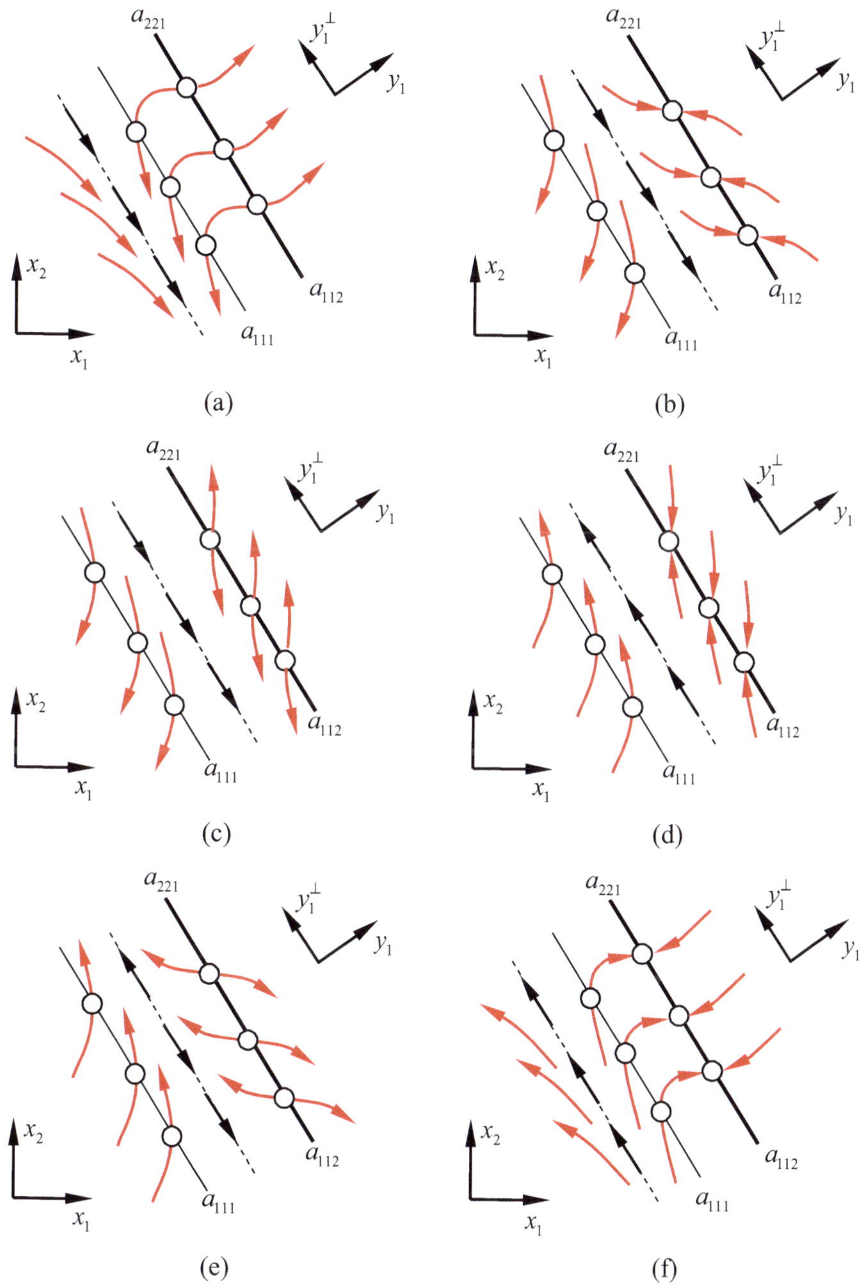

Fig. 3.15 Phase portraits ($a_{111} = a_{221}$ and $\alpha_{11} > 0$, $\alpha_{12} > 0$) for two-dimensional systems on the x_1-direction with $y_1^* = a_{111}, a_{112}$ and on the x_2-direction with $y_1^* = a_{221}$: (**a**) ($a_{110} > 0$, $a_{220} > 0$), (**b**, **c**) ($a_{110} < 0$, $a_{220} > 0$), (**d**, **e**) ($a_{110} > 0$, $a_{220} < 0$), (**f**) ($a_{110} < 0$, $a_{220} < 0$)

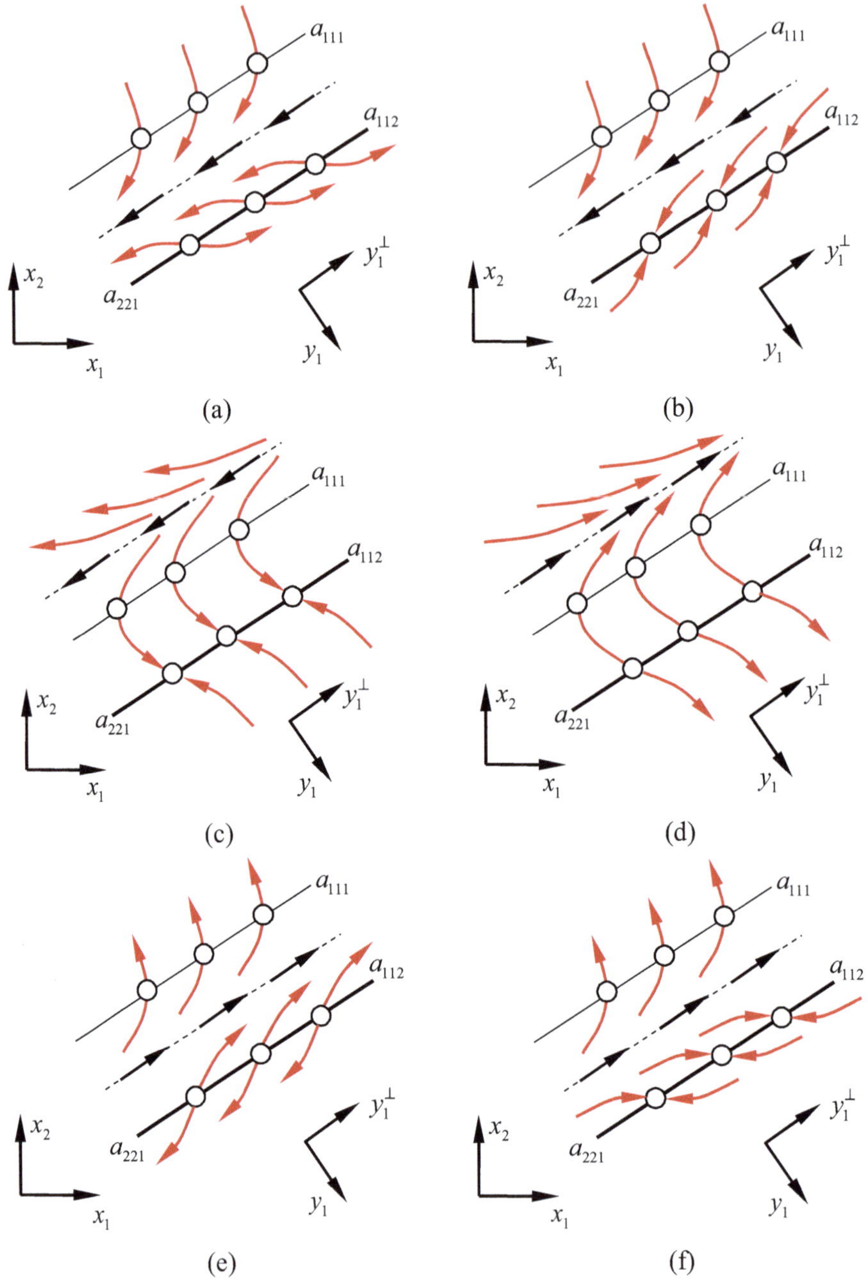

Fig. 3.16 Phase portraits ($a_{112} = a_{221}$ and $\alpha_{11} > 0, \alpha_{12} < 0$) for two-dimensional systems on the x_1-direction with $y_1^* = a_{111}, a_{112}$ and on the x_2-direction with $y_1^* = a_{221}$: (**a**, **b**) ($a_{110} > 0$, $a_{220} > 0$), (**c**) ($a_{110} < 0$, $a_{220} > 0$), (**d**) ($a_{110} > 0$, $a_{220} < 0$), (**e**, **f**) ($a_{110} < 0$, $a_{220} < 0$)

Chapter 4
Single-Linear-Bivariate Quadratic Systems

In this chapter, dynamical systems possessing two quadratic vector fields with a single-linear-bivariate function will be discussed. Two inflection flows exist in such quadratic systems for appearing bifurcations of up- and down-parabola flows. The parabola source and sink infinite-equilibriums are for the switching bifurcations of parabola and inflection flows. The inflection source and sink infinite-equilibriums are for the switching bifurcations of two parabola flows in two directions. The inflection upper and lower-saddle infinite-equilibriums are for the switching bifurcations between two inflection flows in two directions. The corresponding parabola-flow arrays exist in such quadratic systems.

4.1 Parabola and Inflection Flows and Infinite-Equilibriums

In this section, consider dynamical systems with quadratic vector fields with a single-bivariate function. The corresponding stability and bifurcations of equilibriums will be presented through the following theorem.

Theorem 4.1 *Consider a two-dimensional, quadratic dynamical system for* $A_{jj0} \neq 0$ $(j = j_1, j_2 \in \{1, 2\}, j_1 \neq j_2)$ *as*

$$\begin{aligned} \dot{x}_{j_1} &= f_{j_1}(\mathbf{x}, \mathbf{p}) = A_{j_1 j_1 2} + A_{j_1 j_1 1} y_{j_1} + A_{j_1 j_1 0} y_{j_1}^2, \\ \dot{x}_{j_2} &= f_{j_2}(\mathbf{x}, \mathbf{p}) = A_{j_2 j_2 2} + A_{j_2 j_2 1} y_{j_1} + A_{j_2 j_2 0} y_{j_1}^2, \end{aligned} \tag{4.1}$$

with

A. C. J. Luo, *Introduction to Infinite-Equilibriums in Dynamical Systems*,
https://doi.org/10.1007/978-3-031-89083-3_4

$$\begin{aligned}&y_{j_1} = \alpha_{j_1 j_1} x_{j_1} + \alpha_{j_1 j_2} x_{j_2};\\&\Delta_{jj} = (A_{jj1})^2 - 4A_{jj2}A_{jj0}\\&\text{for } j = j_1, j_2 \in \{1, 2\}, j_1 \neq j_2.\end{aligned} \tag{4.2}$$

(i) If

$$\Delta_{j_1 j_1} < 0 \text{ and } \Delta_{j_2 j_2} < 0, \tag{4.3}$$

then

$$\begin{aligned}&\dot{x}_{j_1} = a_{j_1 j_1 0}[(y_{j_1} - a_{j_1 j_1})^2 + b_{j_1 j_1}],\\&\dot{x}_{j_2} = a_{j_2 j_2 0}[(y_{j_1} - a_{j_2 j_2})^2 + b_{j_2 j_2}],\\&\text{with } a_{jj} = \frac{1}{2A_{jj0}} A_{jj1}, b_{jj} = \frac{1}{4A_{jj0}}(-\Delta_{jj}),\\&\text{for } j = j_1, j_2 \in \{1, 2\}; j_1 \neq j_2.\end{aligned} \tag{4.4}$$

- *For* $a_{j_1 j_1 0} > 0$ *and* $a_{j_2 j_2 0} > 0$, *there is a* (pF,pF)-*flow, i.e.,*

$$(\dot{x}_{j_2}, \dot{x}_{j_2}) = \underbrace{(\text{pF}, \text{pF})}_{\text{flow}}. \tag{4.5}$$

- *For* $a_{j_1 j_1 0} < 0$ *and* $a_{j_2 j_2 0} > 0$, *there is an* (nF, pF)-*flow, i.e.,*

$$(\dot{x}_{j_2}, \dot{x}_{j_2}) = \underbrace{(\text{nF}, \text{pF})}_{\text{flow}}. \tag{4.6}$$

- *For* $a_{j_1 j_1 0} > 0$ *and* $a_{j_2 j_2 0} < 0$, *there is a* (pF, nF)-*flow, i.e.,*

$$(\dot{x}_{j_2}, \dot{x}_{j_2}) = \underbrace{(\text{pF}, \text{nF})}_{\text{flow}}. \tag{4.7}$$

- *For* $a_{j_1 j_1 0} < 0$ *and* $a_{j_2 j_2 0} < 0$,*there is an* (nF,nF)-*flow, i.e.,*

$$(\dot{x}_{j_2}, \dot{x}_{j_2}) = \underbrace{(\text{nF}, \text{nF})}_{\text{flow}}. \tag{4.8}$$

(ii) If

$$\Delta_{j_1 j_1} = 0 \text{ and } \Delta_{j_2 j_2} < 0, \tag{4.9}$$

then

$$\begin{aligned}
&\dot{x}_{j_1} = a_{j_1 j_1 0}(y_{j_1} - a_{j_1 j_1 1})^2, \\
&\dot{x}_{j_2} = a_{j_2 j_2 0}[(y_{j_1} - a_{j_2 j_2})^2 + b_{j_2 j_2}], \\
&\text{with } a_{j_1 j_1} = -\frac{1}{2A_{j_1 j_1 0}} A_{j_1 j_1 1}, b_{j_1 j_1} = \frac{1}{2A_{j_1 j_1 0}}(-\Delta_{j_1 j_1}); \\
&a_{j_2 j_2 1} = -\frac{1}{2A_{j_2 j_2 0}} A_{j_2 j_2 1}.
\end{aligned} \tag{4.10}$$

There is a dynamical system in the y_{j_1}-direction *as*

$$\dot{y}_{j_1} = \alpha_{j_1 j_1} a_{j_1 j_1 0}(y_{j_1} - a_{j_1 j_1 1})^2 + \alpha_{j_1 j_2} a_{j_2 j_2 0}[(y_{j_1} - a_{j_2 j_2})^2 + b_{j_2 j_2}], \tag{4.11}$$

with a dynamical system in the $y_{j_1}^{\perp}$*-direction as*

$$\dot{y}_{j_1}^{\perp} = \beta_{j_1 j_1} a_{j_1 j_1 0}(y_{j_1} - a_{j_1 j_1 1})^2 + \beta_{j_1 j_2} a_{j_2 j_2 0}[(y_{j_1} - a_{j_2 j_2})^2 + b_{j_2 j_2}], \tag{4.12}$$

where

$$\begin{aligned}
&y_{j_1}^{\perp} = \beta_{j_1 j_1} x_{j_1} + \beta_{j_1 j_2} x_{j_2}, \mathbf{n}_{j_1}^{\perp} = (\beta_{j_1 j_1}, \beta_{j_1 j_2})^{\mathrm{T}}, \\
&\mathbf{n}_{j_1} \cdot \mathbf{n}_{j_1}^{\perp} = \alpha_{j_1 j_1} \beta_{j_1 j_1} + \alpha_{j_1 j_2} \beta_{j_1 j_2} = 0, \\
&\dot{x}_{j_2}^{*} = a_{j_2 j_2 0}[(a_{j_1 j_1 1} - a_{j_2 j_2 1})^2 + b_{j_2 j_2}].
\end{aligned} \tag{4.13}$$

The equilibrium of $y_{j_1}^{*} = a_{j_1 j_1 1}$ *for the* x_{j_1}*-direction have the following properties with* $\dot{x}_{j_2}^{*} = a_{j_2 j_2 0}[(a_{j_1 j_1 1} - a_{j_2 j_2})^2 + b_{j_2 j_2}]$*:*

- *For* $a_{j_1 j_1 0}\alpha_{j_1 j_2}^2 > 0$ *and* $a_{j_2 j_2 0} > 0$*,*

$$(a_{j_1 j_1 1}, \dot{x}_{j_2}^{*}) = \underbrace{(\mathrm{II}, \mathrm{pF})}_{\text{increasing-inflection flow }(+)}. \tag{4.14}$$

- *For* $a_{j_1 j_1 0}\alpha_{j_1 j_2}^2 < 0$ *and* $a_{j_2 j_2 0} > 0$*,*

$$(a_{j_1 j_1 1}, \dot{x}_{j_2}^{*}) = \underbrace{(\mathrm{DI}, \mathrm{pF})}_{\text{decreasing-inflection flow }(+)}. \tag{4.15}$$

- *For* $a_{j_1 j_1 0}\alpha_{j_1 j_2}^2 > 0$ *and* $a_{j_2 j_2 0} > 0$*,*

$$(a_{j_1 j_1 1}, \dot{x}_{j_2}^{*}) = \underbrace{(\mathrm{DI}, \mathrm{nF})}_{\text{decreasing-inflection flow }(-)}. \tag{4.16}$$

- *For* $a_{j_1j_10}\alpha^2_{j_1j_2} < 0$ *and* $a_{j_2j_20} > 0$,

$$(a_{j_1j_11}, \dot{x}^*_{j_2}) = \underbrace{(\mathrm{II}, \mathrm{nF})}_{\text{increasing-inflection flow }(-)}. \tag{4.17}$$

The increasing-inflection and decreasing-inflection flow bifurcations are for the up-parabola and down-parabola flows.
(iii) If

$$\Delta_{j_1j_1} > 0 \text{ and } \Delta_{j_2j_2} < 0, \tag{4.18}$$

then

$$\begin{aligned} \dot{x}_{j_1} &= a_{j_1j_10}(y_{j_1} - a_{j_1j_11})(y_{j_1} - a_{j_1j_12}), \\ \dot{x}_{j_2} &= a_{j_2j_20}[(y_{j_1} - a_{j_2j_2})^2 + b_{j_2j_2}], \end{aligned} \tag{4.19}$$

where

$$\begin{aligned} &\{a_{j_1j_11}, a_{j_1j_12}\} = \mathrm{sort}\{b_{j_1j_11}, b_{j_1j_12}\}, a_{j_1j_11} < a_{j_1j_12}; \\ &b_{j_1j_11}, b_{j_1j_12} = -\frac{1}{2A_{j_1j_10}}(A_{j_1j_11} \pm \sqrt{\Delta_{j_1j_1}}), a_{j_1j_10} = A_{j_1j_10}. \\ &a_{j_2j_2} = -\frac{1}{2A_{j_2j_20}}A_{j_2j_21}, b_{j_2j_2} = \frac{1}{4A_{j_2j_20}}(-\Delta_{j_2j_2}), j_2 \in \{1, 2\}. \end{aligned} \tag{4.20}$$

There is a dynamical system in the y_{j_1}-direction *as*

$$\begin{aligned} \dot{y}_{j_1} &= \alpha_{j_1j_1}a_{j_1j_10}(y_{j_1} - a_{j_1j_11})(y_{j_1} - a_{j_1j_12}) \\ &\quad + \alpha_{j_1j_2}a_{j_2j_20}[(y_{j_1} - a_{j_2j_2})^2 + b_{j_2j_2}], \end{aligned} \tag{4.21}$$

with a dynamical system in the $y^{\perp}_{j_1}$*-direction as*

$$\begin{aligned} \dot{y}^{\perp}_{j_1} &= \beta_{j_1j_1}a_{j_1j_10}(y_{j_1} - a_{j_1j_11})(y_{j_1} - a_{j_1j_12}) \\ &\quad + \beta_{j_1j_2}a_{j_2j_20}[(y_{j_1} - a_{j_2j_2})^2 + b_{j_2j_2}]. \end{aligned} \tag{4.22}$$

The equilibrium of $y^*_{j_1} = a_{j_1j_1s}$*(*$s, l \in \{1, 2\}, s \neq l$*) in the* x_{j_1}*-direction possesses the following properties:*

- *For* $a_{j_1j_10}\alpha_{j_1j_2}(a_{j_1j_1s} - a_{j_1j_1l}) > 0$ *and* $a_{j_2j_20} > 0$,

$$(a_{j_1j_1s}, \dot{x}^*_{j_2}) = \underbrace{(\mathrm{UP}, \mathrm{pF})}_{\text{up-parabola flow }(+)}. \tag{4.23}$$

- *For* $a_{j_1j_10}\alpha_{j_1j_2}(a_{j_1j_1s}-a_{j_1j_1l})<0$ *and* $a_{j_2j_20}>0$,

$$(a_{j_1j_1s},\dot{x}^*_{j_2}) = \underbrace{(\mathrm{DP},\mathrm{pF})}_{\text{down-parabola flow }(+)}. \tag{4.24}$$

- *For* $a_{j_1j_10}\alpha_{j_1j_2}(a_{j_1j_1s}-a_{j_1j_1l})>0$ *and* $a_{j_2j_20}<0$,

$$(a_{j_1j_1s},\dot{x}^*_{j_2}) = \underbrace{(\mathrm{DP},\mathrm{nF})}_{\text{down-parabola flow }(-)}. \tag{4.25}$$

- *For* $a_{j_1j_10}\alpha_{j_1j_2}(a_{j_1j_1s}-a_{j_1j_1l})<0$ *and* $a_{j_2j_20}<0$,

$$(a_{j_1j_1s},\dot{x}^*_{j_2}) = \underbrace{(\mathrm{UP},\mathrm{nF})}_{\text{up-parabola flow }(-)}. \tag{4.26}$$

(iv) If

$$\Delta_{j_1j_1}>0 \text{ and } \Delta_{j_2j_2}=0, \tag{4.27}$$

then

$$\begin{aligned}
\dot{x}_{j_1} &= a_{j_1j_10}(y_{j_1}-a_{j_1j_11})(y_{j_1}-a_{j_1j_12}),\\
\dot{x}_{j_2} &= a_{j_2j_20}(y_{j_1}-a_{j_2j_21})^2,
\end{aligned} \tag{4.28}$$

where

$$\begin{aligned}
&\{a_{j_1j_11},a_{j_1j_12}\} = \mathrm{sort}\{b_{j_1j_11},b_{j_1j_12}\},\, a_{j_1j_11}<a_{j_1j_12};\\
&b_{j_1j_11},b_{j_1j_12} = -\frac{1}{2A_{j_1j_10}}\left(A_{j_1j_11}\pm\sqrt{\Delta_{j_1j_1}}\right),\, a_{j_1j_10}=A_{j_1j_10}.\\
&a_{j_2j_21} = -\frac{1}{2A_{j_2j_20}}A_{j_2j_21}, j_2\in\{1,2\}.
\end{aligned} \tag{4.29}$$

There is a dynamical system in the y_{j_1}*-direction as*

$$\dot{y}_{j_1} = \alpha_{j_1j_1}a_{j_1j_10}(y_{j_1}-a_{j_1j_11})(y_{j_1}-a_{j_1j_12}) + \alpha_{j_1j_2}a_{j_2j_20}(y_{j_1}-a_{j_2j_21})^2, \tag{4.30}$$

with a dynamical system in the $y^{\perp}_{j_1}$*-direction as*

$$\dot{y}^{\perp}_{j_1} = \beta_{j_1j_1}a_{j_1j_10}(y_{j_1}-a_{j_1j_11})(y_{j_1}-a_{j_1j_12}) + \beta_{j_1j_2}a_{j_2j_20}(y_{j_1}-a_{j_2j_21})^2. \tag{4.31}$$

(iv$_1$) The simple equilibriums of $y^*_{j_1} = a_{j_1 j_1 s}$*(*$s,\ l \in \{1,2\},\ s \neq l$*) for the* x_{j_1}*-direction have the following properties with* $\dot{x}^*_{j_2} = a_{j_2 j_2 0}(a_{j_1 j_1 s} - a_{j_2 j_2 1})^2$*:*

- *For* $a_{j_1 j_1 0}\alpha_{j_1 j_2}(a_{j_1 j_1 s} - a_{j_1 j_1 l}) > 0$ *and* $a_{j_2 j_2 0}(a_{j_1 j_1 s} - a_{j_2 j_2 1})^2 > 0$,

$$(a_{j_1 j_1 s}, \dot{x}^*_{j_2}) = \underbrace{(\mathrm{UP}, \mathrm{pF})}_{\text{up-parabola flow } (+)}. \tag{4.32}$$

- *For* $a_{j_1 j_1 0}\alpha_{j_1 j_2}(a_{j_1 j_1 s} - a_{j_1 j_1 l}) < 0$ *and* $a_{j_2 j_2 0}(a_{j_1 j_1 s} - a_{j_2 j_2 1})^2 > 0$,

$$(a_{j_1 j_1 s}, \dot{x}^*_{j_2}) = \underbrace{(\mathrm{DP}, \mathrm{pF})}_{\text{Down-parabola flow } (+)}. \tag{4.33}$$

- *For* $a_{j_1 j_1 0}\alpha_{j_1 j_2}(a_{j_1 j_1 s} - a_{j_1 j_1 l}) > 0$ *and* $a_{j_2 j_2 0}(a_{j_1 j_1 s} - a_{j_2 j_2 1})^2 < 0$,

$$(a_{j_1 j_1 s}, \dot{x}^*_{j_2}) = \underbrace{(\mathrm{DP}, \mathrm{nF})}_{\text{Down-parabola flow } (-)}. \tag{4.34}$$

- *For* $a_{j_1 j_1 0}\alpha_{j_1 j_2}(a_{j_1 j_1 s} - a_{j_1 j_1 l}) < 0$ *and* $a_{j_2 j_2 0}(a_{j_1 j_1 s} - a_{j_2 j_2 1})^2 < 0$,

$$(a_{j_1 j_1 s}, \dot{x}^*_{j_2}) = \underbrace{(\mathrm{UP}, \mathrm{nF})}_{\text{up-parabola flow } (-)}. \tag{4.35}$$

(iv$_2$) The repeated equilibrium of $y^*_{j_1} = a_{j_2 j_2 1}$ *for the* x_{j_2}*-direction has the following properties with* $\dot{x}^*_{j_1} = a_{j_1 j_1 0}\prod_{s=1}^{2}(a_{j_2 j_2 1} - a_{j_1 j_1 s})$*:*

- For $a_{j_1 j_1 0}\prod_{s=1}^{2}(a_{j_2 j_2 1} - a_{j_1 j_1 s}) > 0$ and $a_{j_2 j_2 0} > 0$,

$$(\dot{x}^*_{j_1}, a_{j_2 j_2 1}) = \underbrace{(\mathrm{pF}, \mathrm{II})}_{\text{increaseing-inflection flow } (+)}. \tag{4.36}$$

- For $a_{j_1 j_1 0}\prod_{s=1}^{2}(a_{j_2 j_2 1} - a_{j_1 j_1 s}) < 0$ and $a_{j_2 j_2 0} > 0$,

$$(\dot{x}^*_{j_1}, a_{j_2 j_2 1}) = \underbrace{(\mathrm{nF}, \mathrm{DI})}_{\text{decreaseing-inflection flow } (-)}. \tag{4.37}$$

- For $a_{j_1 j_1 0}\prod_{s=1}^{2}(a_{j_2 j_2 1} - a_{j_1 j_1 s}) > 0$ and $a_{j_2 j_2 0} < 0$,

$$(\dot{x}^*_{j_1}, a_{j_2 j_2 1}) = \underbrace{(\mathrm{pF}, \mathrm{DI})}_{\text{decreaseing-inflection flow } (+)}. \tag{4.38}$$

- For $a_{j_1j_10}\prod_{s=1}^{2}(a_{j_2j_21}-a_{j_1j_1s})<0$ and $a_{j_2j_20}<0$,

$$(\dot{x}_{j_1}^{*},a_{j_2j_21})=\underbrace{(\mathrm{nF},\mathrm{II})}_{\text{increaseing-inflection flow }(-)}. \tag{4.39}$$

The decreasing- and increasing-inflection flow are the appearing bifurcations for the up-parabola and parabola flows.

(iv$_3$) For $a_{j_2j_21}=a_{j_1j_1s}$ $(s\in\{1,2\})$

$$\begin{aligned}&\dot{x}_{j_1}=a_{j_1j_10}(y_{j_1}-a_{j_1j_1s})(y_{j_1}-a_{j_1j_1l}),\\&\dot{x}_{j_2}=a_{j_2j_20}(y_{j_1}-a_{j_2j_21})^{2},\\&j_1,j_2\in\{1,2\};j_1\neq j_2;s,l\in\{1,2\},s\neq l.\end{aligned} \tag{4.40}$$

There is a dynamical system in the y_{j_1}-direction *as*

$$\dot{y}_{j_1}=b_{j_1j_10}(y_{j_1}-a_{j_1j_1s})(y_{j_1}-b_{j_1j_1l}), \tag{4.41}$$

with a dynamical system in the $y_{j_1}^{\perp}$*-direction as*

$$\dot{y}_{j_1}^{\perp}=b_{j_1j_10}^{\perp}(y_{j_1}-a_{j_1j_1s})(y_{j_1}-b_{j_1j_1l}^{\perp}), \tag{4.42}$$

where

$$\begin{aligned}&b_{j_1j_10}=(\alpha_{j_1j_1}a_{j_1j_10}+\alpha_{j_1j_1}a_{j_1j_10}),b_{j_1j_1l}=\frac{\alpha_{j_1j_1}a_{j_1j_12}+\alpha_{j_1j_2}a_{j_1j_11}}{\alpha_{j_1j_1}a_{j_1j_10}+\alpha_{j_1j_1}a_{j_1j_10}},\\&b_{j_1j_10}^{\perp}=(\beta_{j_1j_1}a_{j_1j_10}+\beta_{j_1j_1}a_{j_1j_10}),b_{j_1j_1l}^{\perp}=\frac{\beta_{j_1j_1}a_{j_1j_12}+\beta_{j_1j_2}a_{j_1j_11}}{\beta_{j_1j_1}a_{j_1j_10}+\beta_{j_1j_1}a_{j_1j_10}}.\end{aligned} \tag{4.43}$$

The infinite-equilibriums of $y_{j_1}^{*}=a_{j_1j_1s}=a_{j_2j_21}$ *for the* x_{j_1}*-direction have the following properties:*

- *For* $\alpha_{j_1j_1}a_{j_1j_10}(a_{j_1j_1s}-a_{j_2j_2l})>0$ *and* $a_{j_2j_20}\alpha_{j_1j_1}^{2}>0$,

$$(a_{j_1j_1s},a_{j_2j_21})=\underbrace{(\mathrm{SO},\mathrm{UP})}_{\text{infinite-equilibrium}}. \tag{4.44}$$

- *For* $\alpha_{j_1j_1}a_{j_1j_10}(a_{j_1j_1s}-a_{j_2j_2l})<0$ *and* $a_{j_2j_20}\alpha_{j_1j_1}^{2}>0$,

$$(a_{j_1j_1s},a_{j_2j_21})=\underbrace{(\mathrm{SI},\mathrm{DP})}_{\text{infinite-equilibrium}}. \tag{4.45}$$

- *For* $\alpha_{j_1j_1}a_{j_1j_10}(a_{j_1j_1s}-a_{j_2j_2l})>0$ *and* $a_{j_2j_20}\alpha_{j_1j_1}^2<0$,

$$(a_{j_1j_1s},a_{j_2j_21})=\underbrace{(\mathrm{SO},\mathrm{DP})}_{\text{infinite-equilibrium}}. \tag{4.46}$$

- *For* $\alpha_{j_1j_1}a_{j_1j_10}(a_{j_1j_1s}-a_{j_2j_2l})<0$ *and* $a_{j_2j_20}\alpha_{j_1j_1}^2<0$,

$$(a_{j_1j_1s},a_{j_2j_21})=\underbrace{(\mathrm{SI},\mathrm{UP})}_{\text{infinite-equilibrium}}. \tag{4.47}$$

(v) If

$$\Delta_{j_1j}>0 \text{ and } \Delta_{j_2j}>0, \tag{4.48}$$

then

$$\begin{aligned}
\dot{x}_{j_1}&=a_{j_1j_10}(y_{j_1}-a_{j_1j_11})(y_{j_1}-a_{j_1j_12}),\\
\dot{x}_{j_2}&=a_{j_2j_20}(y_{j_1}-a_{j_2j_21})(y_{j_1}-a_{j_2j_22}),\\
&j_1,j_2\in\{1,2\};j_1\neq j_2.
\end{aligned} \tag{4.49}$$

where

$$\begin{aligned}
&\{a_{jj1},a_{jj2}\}=\mathrm{sort}\{b_{jj1},b_{jj2}\},a_{jj1}<a_{jj2};\\
&b_{jj1},b_{jj2}=-\frac{1}{2A_{jj0}}(A_{jj1}\pm\sqrt{\Delta_{jj}}),a_{jj0}=A_{jj0};\\
&j=j_1,j_2\in\{1,2\};j_1\neq j_2.
\end{aligned} \tag{4.50}$$

There is a dynamical system in the y_{j_1}-direction *as*

$$\begin{aligned}
\dot{y}_{j_1}&=\alpha_{j_1j_1}a_{j_1j_10}(y_{j_1}-a_{j_1j_11})(y_{j_1}-a_{j_1j_12})\\
&+\alpha_{j_1j_2}a_{j_2j_20}(y_{j_1}-a_{j_2j_21})(y_{j_1}-a_{j_2j_22}),
\end{aligned} \tag{4.51}$$

with a dynamical system in the $y_{j_1}^{\perp}$*-direction as*

$$\begin{aligned}
\dot{y}_{j_1}^{\perp}&=\beta_{j_1j_1}a_{j_1j_10}(y_{j_1}-a_{j_1j_11})(y_{j_1}-a_{j_1j_12})\\
&+\beta_{j_1j_2}a_{j_2j_20}(y_{j_1}-a_{j_2j_21})(y_{j_1}-a_{j_2j_22}).
\end{aligned} \tag{4.52}$$

(v_1) The equilibrium of $y_{j_1}^*=a_{j_1j_1s_1},a_{j_1j_1s_2}$ $(s_1,s_2\in\{1,2\},\ s_2\neq s_1)$ *for the* x_{j_1}*-direction has four cases with* $\dot{x}_{j_2}^*=a_{j_2j_20}\prod_{l_1=1}^{2}(a_{j_1j_1s_1}-a_{j_2j_2l_1})$.

- *For* $\alpha_{j_1j_2}a_{j_1j_1 0}(a_{j_1j_2s_1}-a_{j_1j_2s_2})>0$ *and* $a_{j_2j_2 0}\prod_{l_1=1}^{2}(a_{j_1j_1s_1}-a_{j_2j_2l_1})>0$,

$$(a_{j_1j_1s_1},\dot{x}_{j_2}^{*})=\underbrace{(\mathrm{UP},\mathrm{pF})}_{\text{up-parabola flow }(+)}. \tag{4.53}$$

- *For* $\alpha_{j_1j_2}a_{j_1j_1 0}(a_{j_1j_2s_1}-a_{j_1j_2s_2})<0$ *and* $a_{j_2j_2 0}\prod_{l_1=1}^{2}(a_{j_1j_1s_1}-a_{j_2j_2l_1})>0$,

$$(a_{j_1j_1s_1},\dot{x}_{j_2}^{*})=\underbrace{(\mathrm{DP},\mathrm{pF})}_{\text{down-parabola flow }(+)}. \tag{4.54}$$

- *For* $\alpha_{j_1j_2}a_{j_1j_1 0}(a_{j_1j_2s_1}-a_{j_1j_2s_2})>0$ *and* $a_{j_2j_2 0}\prod_{l_1=1}^{2}(a_{j_1j_1s_1}-a_{j_2j_2l_1})<0$,

$$(a_{j_1j_1s_1},\dot{x}_{j_2}^{*})=\underbrace{(\mathrm{DP},\mathrm{nF})}_{\text{down-parabola flow }(-)}. \tag{4.55}$$

- *For* $\alpha_{j_1j_2}a_{j_1j_1 0}(a_{j_1j_2s_1}-a_{j_1j_2s_2})<0$ *and* $a_{j_2j_2 0}\prod_{l_1=1}^{2}(a_{j_1j_1s_1}-a_{j_2j_2l_1})<0$,

$$(a_{j_1j_1,s_1},\dot{x}_{j_2}^{*})=\underbrace{(\mathrm{UP},\mathrm{nF})}_{\text{up-parabola flow }(-)}. \tag{4.56}$$

(v_2*) The equilibrium of* $y_{j_1}^{*}=a_{j_2j_2l_1}$ *(*$l_1,\ l_2\in\{1,2\},\ l_1\neq l_2$*) for the* x_{j_2}*-direction has four cases with* $\dot{x}_{j_1}^{*}=a_{j_1j_1 0}\prod_{s_1=1}^{2}(a_{j_2j_2l_1}-a_{j_1j_1s_1})$.

- *For* $a_{j_1j_1 0}\prod_{s_1=1}^{2}(a_{j_2j_2l_1}-a_{j_1j_1s_1})>0$ *and* $\alpha_{j_1j_1}a_{j_2j_2 0}(a_{j_2j_2l_1}-a_{j_2j_2l_2})>0$,

$$(\dot{x}_{j_1}^{*},a_{j_2j_2l_1})=\underbrace{(\mathrm{pF},\mathrm{UP})}_{\text{up-parabola flow }(+)}. \tag{4.57}$$

- *For* $a_{j_1j_1 0}\prod_{s_1=1}^{2}(a_{j_2j_2l_1}-a_{j_1j_1s_1})<0$ *and* $\alpha_{j_1j_1}a_{j_2j_2 0}(a_{j_2j_2l_1}-a_{j_2j_2l_2})>0$,

$$(\dot{x}_{j_1}^{*},a_{j_2j_2l_1})=\underbrace{(\mathrm{nF},\mathrm{DP})}_{\text{down-parabola flow }(-)}. \tag{4.58}$$

- *For* $a_{j_1j_1 0}\prod_{s_1=1}^{2}(a_{j_2j_2l_1}-a_{j_1j_1s_1})>0$ *and* $\alpha_{j_1j_1}a_{j_2j_2 0}(a_{j_2j_2l_1}-a_{j_2j_2l_2})<0$,

$$(\dot{x}_{j_1}^{*},a_{j_2j_2l_1})=\underbrace{(\mathrm{pF},\mathrm{DP})}_{\text{down-parabola flow }(+)}. \tag{4.59}$$

- *For* $a_{j_1j_10}\prod_{s_1=1}^{2}(a_{j_2j_2l_1}-a_{j_1j_1s_1})<0$ *and* $\alpha_{j_1j_1}a_{j_2j_20}(a_{j_2j_2l_1}-a_{j_2j_2l_2})<0$,

$$(\dot{x}_{j_1}^{*}, a_{j_2j_2l_1}) = \underbrace{(\mathrm{nF}, \mathrm{UP})}_{\text{up-parabola flow }(-)}. \tag{4.60}$$

(v_3) For $a_{j_1j_1s_1}=a_{j_1j_1l_1}(s_1,\, l_1\in\{1,2\})$,

$$\begin{aligned}
&\dot{x}_{j_1}=a_{j_1j_10}(y_{j_1}-a_{j_1js_1})(y_{j_1}-a_{j_1js_2}),\\
&\dot{x}_{j_2}=a_{j_2j_20}(y_{j_1}-a_{j_2j_2l_1})(y_{j_1}-a_{j_2j_2l_2}),\\
&j_1,j_2\in\{1,2\};j_1\neq j_2,\\
&s_1,s_2,l_1,l_2\in\{1,2\};s_1\neq s_2,l_1\neq l_2.
\end{aligned}\tag{4.61}$$

There is a dynamical system in the y_{j_1}-direction *as*

$$\dot{y}_{j_1}=(y_{j_1}-a_{j_1j_1s_1})[\alpha_{j_1j_1}a_{j_1j_10}(y_{j_1}-a_{j_1j_1s_2})+\alpha_{j_1j_2}a_{j_2j_20}(y_{j_1}-a_{j_2j_2l_2})], \tag{4.62}$$

with a dynamical system in the $y_{j_1}^{\perp}$*-direction as*

$$\dot{y}_{j_1}^{\perp}=(y_{j_1}-a_{j_1j_1s_1})[\beta_{j_1j_1}a_{j_1j_10}(y_{j_1}-a_{j_1j_1s_2})+\beta_{j_1j_2}a_{j_2j_20}(y_{j_1}-a_{j_2j_2l_2})], \tag{4.63}$$

where

$$G_{j_1j_10}=\alpha_{j_1j_1}a_{j_1j_10}(a_{j_1j_1s_1}-a_{j_1j_1s_2})+\alpha_{j_1j_2}a_{j_2j_20}(a_{j_2j_2l_1}-a_{j_2j_2l_2}). \tag{4.64}$$

The infinite-equilibrium of $y_{j_1}^{*}=a_{j_1j_1s_1}=a_{j_2j_2l_1}$ *with* $a_{j_1j_1s_2}\neq a_{j_2j_2l_2}$ *has the following properties:*

- *For* $a_{j_1j_10}(a_{j_1j_1s_1}-a_{j_1j_1s_2})/a_{j_2j_20}(a_{j_2j_2l_1}-a_{j_2j_2l_2})>0$ *and* $G_{j_1j_10}>0$,

$$(a_{j_1j_1s_1}, a_{j_2j_2l_1}) = \underbrace{(\mathrm{II}, \mathrm{SO})}_{\text{increasing-inflection source}}. \tag{4.65}$$

- *For* $a_{j_1j_10}(a_{j_1j_1s_1}-a_{j_1j_1s_2})/a_{j_2j_20}(a_{j_2j_2l_1}-a_{j_2j_2l_2})>0$ *and* $G_{j_1j_10}>0$,

$$(a_{j_1j_1s_1}, a_{j_2j_2l_1}) = \underbrace{(\mathrm{II}, \mathrm{SI})}_{\text{increasing-inflection sink}}. \tag{4.66}$$

- *For* $a_{j_1j_10}(a_{j_1j_1s_1}-a_{j_1j_1s_2})/a_{j_2j_20}(a_{j_2j_2l_1}-a_{j_2j_2l_2})<0$ *and* $G_{j_1j_10}>0$,

$$(a_{j_1j_1s_1}, a_{j_2j_2l_1}) = \underbrace{(\mathrm{DI}, \mathrm{SO})}_{\text{decreasing-inflection source}}. \tag{4.67}$$

- *For* $a_{j_1j_10}(a_{j_1j_1s_1} - a_{j_1j_1s_2})/a_{j_2j_20}(a_{j_2j_2l_1} - a_{j_2j_2l_2}) < 0$ *and* $G_{j_1j_10} < 0$,

$$(a_{j_1j_1s_1}, a_{j_2j_2l_1}) = \underbrace{(\mathrm{DI}, \mathrm{SI})}_{\text{decreasing-inflection sink}}. \tag{4.68}$$

The infinite-equilibriums are the switching bifurcations for the down-parabola and up-parabola flows with the up-parabola and down-parabola flows.

(v_4) For $a_{j_1j_11} = a_{j_2j_21}$ *and* $a_{j_1j_12} = a_{j_2j_22}$ *but* $a_{j_1j_11} \neq a_{j_1j_12}$

$$\begin{aligned}
&\dot{x}_{j_1} = a_{j_1j_10}(y_{j_1} - a_{j_1j_11})(y_{j_1} - a_{j_1j_12}),\\
&\dot{x}_{j_2} = a_{j_2j_20}(y_{j_1} - a_{j_2j_21})(y_{j_1} - a_{j_2j_22}),\\
&j_1, j_2 \in \{1,2\}; j_1 \neq j_2.
\end{aligned} \tag{4.69}$$

There is a dynamical system in the y_{j_1}-direction *as*

$$\dot{y}_{j_1} = (\alpha_{j_1j_1}a_{j_1j_10} + \alpha_{j_1j_2}a_{j_2j_20})(y_{j_1} - a_{j_1j_11})(y_{j_1} - a_{j_1j_12}) \tag{4.70}$$

with a dynamical system in the $y_{j_1}^{\perp}$*-direction as*

$$\dot{y}_{j_1}^{\perp} = (\beta_{j_1j_1}a_{j_1j_10} + \beta_{j_1j_2}a_{j_2j_20})(y_{j_1} - a_{j_1j_11})(y_{j_1} - a_{j_1j_12}), \tag{4.71}$$

and

$$G_{j_1j_10} = (\alpha_{j_1j_1}a_{j_1j_10} + \alpha_{j_1j_2}a_{j_2j_20})(a_{j_2j_2l_1} - a_{j_2j_2l_2}). \tag{4.72}$$

The infinite-equilibrium of $a_{j_1j_1s_1} = a_{j_2j_2l_1}$ *and* $a_{j_1j_1s_2} = a_{j_2j_2l_2}$ *with* $a_{j_1j_1s_1} \neq a_{j_1j_1s_2}$ (s_1, $s_2 \in \{1,2\}, s_1 \neq s_2$ and l_1, $l_2 \in \{1,2\}, l_1 \neq l_2$) *have four cases:*

- *For* $a_{j_1j_10}/a_{j_2j_20} > 0$ *and* $G_{j_1j_10} > 0$,

$$(a_{j_1j_1s_1}, a_{j_2j_2l_1}) = \underbrace{(\mathrm{II}, \mathrm{SO})}_{\text{increasing-inflection source}}. \tag{4.73}$$

- *For* $a_{j_1j_10}/a_{j_2j_20} > 0$ *and* $G_{j_1j_10} > 0$,

$$(a_{j_1j_1s_1}, a_{j_2j_2l_1}) = \underbrace{(\mathrm{II}, \mathrm{SI})}_{\text{increasing-inflection sink}}. \tag{4.74}$$

- *For* $a_{j_1j_10}/a_{j_2j_20} < 0$ *and* $G_{j_1j_10} > 0$,

$$(a_{j_1j_1s_1}, a_{j_2j_2l_1}) = \underbrace{(\mathrm{DI}, \mathrm{SO})}_{\text{decreasing-inflection source}}. \tag{4.75}$$

- *For* $a_{j_1j_10}/a_{j_2j_20} < 0$ *and* $G_{j_1j_10} < 0$,

$$(a_{j_1j_1s_1}, a_{j_2j_2l_1}) = \underbrace{(\text{DI}, \text{SI})}_{\text{decreasing-inflection sink}} . \tag{4.76}$$

(vi) Consider a dynamical system with $a_{j_1j_11} \neq a_{j_2j_21}$

$$\begin{aligned} \dot{x}_{j_1} &= a_{j_1j_10}(y_{j_1} - a_{j_1j_11})^2, \\ \dot{x}_{j_2} &= a_{j_2j_20}(y_{j_1} - a_{j_2j_21})^2, \\ j_1, j_2 &\in \{1,2\}; j_1 \neq j_2. \end{aligned} \tag{4.77}$$

There is a dynamical system in the y_{j_1}-direction *as*

$$\dot{y}_{j_1} = \alpha_{j_1j_1}a_{j_1j_10}(y_{j_1} - a_{j_1j_11})^2 + \alpha_{j_1j_2}a_{j_2j_20}(y_{j_1} - a_{j_2j_21})^2, \tag{4.78}$$

with a dynamical system in the $y_{j_1}^{\perp}$*-direction as*

$$\dot{y}_{j_1}^{\perp} = \beta_{j_1j_1}a_{j_1j_10}(y_{j_1} - a_{j_1j_11})^2 + \beta_{j_1j_2}a_{j_2j_20}(y_{j_1} - a_{j_2j_21})^2. \tag{4.79}$$

(vi$_1$) The flow of $y_{j_1}^* = a_{j_1j_11}$ *for the* x_{j_1}*-direction with* $a_{j_1j_11} \neq a_{j_2j_21}$ *have the following properties:*

- *For* $a_{j_1j_10} > 0$ *and* $a_{j_2j_20}(a_{j_1j_11} - a_{j_2j_21})^2 > 0$,

$$(a_{j_1j_11}, \dot{x}_{j_2}^*) = \underbrace{(\text{II}, \text{pF})}_{\text{increasing-inflection flow }(+)} . \tag{4.80}$$

- *For* $a_{j_1j_10} < 0$ *and* $a_{j_2j_20}(a_{j_1j_11} - a_{j_2j_21})^2 > 0$,

$$(a_{j_1j_11}, \dot{x}_{j_2}^*) = \underbrace{(\text{DI}, \text{pF})}_{\text{derecasing-inflection flow }(+)} . \tag{4.81}$$

- *For* $a_{j_1j_10} > 0$ *and* $a_{j_2j_20}(a_{j_1j_11} - a_{j_2j_21})^2 < 0$,

$$(a_{j_1j_11}, \dot{x}_{j_2}^*) = \underbrace{(\text{DI}, \text{nF})}_{\text{decreasing-inflection flow }(-)} . \tag{4.82}$$

- *For* $a_{j_1j_10} > 0$ *and* $a_{j_2j_20}(a_{j_1j_11} - a_{j_2j_21})^2 < 0$,

$$(a_{j_1j_11}, \dot{x}_{j_2}^*) = \underbrace{(\text{II}, \text{nF})}_{\text{increasing-inflection flow }(-)} . \tag{4.83}$$

(vi_2) The flow of $y^*_{j_1} = a_{j_2j_2 1}$ for the x_{j_1}-direction with $a_{j_1j_1 1} \neq a_{j_2j_2 1}$ have the following properties:

- *For $a_{j_1j_1 0}(a_{j_2j_2 1} - a_{j_1j_1 1})^2 > 0$ and $a_{j_2j_2 0} > 0$,*

$$(\dot{x}^*_{j_1}, a_{j_2j_2 1}) = \underbrace{(\text{pF}, \text{II})}_{\text{increasing-inflection flow } (+)}. \tag{4.84}$$

- *For $a_{j_1j_1 0}(a_{j_2j_2 1} - a_{j_1j_1 1})^2 < 0$ and $a_{j_2j_2 0} > 0$,*

$$(\dot{x}^*_{j_1}, a_{j_2j_2 1}) = \underbrace{(\text{nF}, \text{DI})}_{\text{decreasing-inflection flow } (-)}. \tag{4.85}$$

- *For $a_{j_1j_1 0}(a_{j_2j_2 1} - a_{j_1j_1 1})^2 > 0$ and $a_{j_2j_2 0} < 0$,*

$$(\dot{x}^*_{j_1}, a_{j_2j_2 1}) = \underbrace{(\text{pF}, \text{DI})}_{\text{decreasing-inflection flow } (+)}. \tag{4.86}$$

- *For $a_{j_1j_1 0}(a_{j_2j_2 1} - a_{j_1j_1 1})^2 < 0$ and $a_{j_2j_2 0} > 0$,*

$$(\dot{x}^*_{j_1}, a_{j_2j_2 1}) = \underbrace{(\text{nF}, \text{II})}_{\text{increasing-inflection flow } (-)}. \tag{4.87}$$

The increasing and decreasing inflection flows are the appearing bifurcations of for the up-parabola and down-parabola flows.

(vi_3) Consider a dynamical system with $a_{j_1j_1 1} = a_{j_2j_2 1}$,

$$\begin{aligned}
&\dot{x}_{j_1} = a_{j_1j_1 0}(y_{j_1} - a_{j_1j_1 1})^2,\\
&\dot{x}_{j_2} = a_{j_2j_2 0}(y_{j_1} - a_{j_2j_2 1})^2,\\
&j_1, j_2 \in \{1, 2\}; j_1 \neq j_2.
\end{aligned} \tag{4.88}$$

There is a dynamical system in the y_{j_1}-direction *as*

$$\dot{y}_{j_1} = (\alpha_{j_1j_1} a_{j_1j_1 0} + \alpha_{j_1j_2} a_{j_2j_2 0})(y_{j_1} - a_{j_1j_1 1})^2, \tag{4.89}$$

with a dynamical system in the $y^{\perp}_{j_1}$-direction as

$$\dot{y}^{\perp}_{j_1} = (\beta_{j_1j_1} a_{j_1j_1 0} + \beta_{j_1j_2} a_{j_2j_2 0})(y_{j_1} - a_{j_1j_1 1})^2, \tag{4.90}$$

and

$$G^{(1)}_{j_1 j_1 0} = \alpha_{j_1 j_1} a_{j_1 j_1 0} + \alpha_{j_1 j_2} a_{j_2 j_2 0}. \tag{4.91}$$

The infinite-equilibrium of $y^*_{j_1} = a_{j_1 j_1 1} = a_{j_2 j_2 1}$ *has the following properties:*

- *For* $a_{j_1 j_1 0}/a_{j_2 j_2 0} > 0$ *and* $G^{(1)}_{j_1 j_1 0} > 0$,

$$(a_{j_1 j_1 1}, a_{j_2 j_2 1}) = \underbrace{(\mathrm{II}, \mathrm{US})}_{\text{increasing-inflection upper-saddle}}. \tag{4.92}$$

- *For* $a_{j_1 j_1 0}/a_{j_2 j_2 0} > 0$ *and* $G^{(1)}_{j_1 j_1 0} > 0$,

$$(a_{j_1 j_1 1}, a_{j_2 j_2 1}) = \underbrace{(\mathrm{II}, \mathrm{LS})}_{\text{increasing-inflection lower-saddle}}. \tag{4.93}$$

- *For* $a_{j_1 j_1 0}/a_{j_2 j_2 0} < 0$ *and* $G^{(1)}_{j_1 j_1 0} > 0$,

$$(a_{j_1 j_1 1}, a_{j_2 j_2 1}) = \underbrace{(\mathrm{DI}, \mathrm{US})}_{\text{decreasing-inflection upper-saddle}}. \tag{4.94}$$

- *For* $a_{j_1 j_1 0}/a_{j_2 j_2 0} < 0$ *and* $G^{(1)}_{j_1 j_1 0} > 0$,

$$(a_{j_1 j_1 1}, a_{j_2 j_2 1}) = \underbrace{(\mathrm{DI}, \mathrm{LS})}_{\text{decreasing-inflection lower-saddle}}. \tag{4.95}$$

The infinite-equilibriums are the switching bifurcations for the up-parabola and down-parabola flows in the x_{j_1} *and* x_{j_2}*-directions.*

4.2 Proof of Theorem 4.1

(i) From differential equations for dynamical systems, the results are obtained straightforward.
(ii) From

$$\begin{aligned} \dot{x}_{j_1} &= a_{j_1 j_1 0}(y_{j_1} - a_{j_1 j_1 1})^2, \\ \dot{x}_{j_2} &= a_{j_2 j_2 0}[(y_{j_1} - a_{j_2 j_2 1})^2 + b_{j_2 j_2}], \end{aligned}$$

the dynamical system in the y_{j_1} -direction and the $y^{\perp}_{j_1}$ -direction can be easily constructed in Eqs. (4.11)–(4.13).

From the differential equation, for $y^*_{j_1} = a_{j_1 j_1 1}$,

$$\frac{dx_{j_1}}{dx_{j_2}}\Big|_{y^*_{j_1}=a_{j_1j_1 1}} = \frac{a_{j_1j_1 0}(y_{j_1}-a_{j_1j_1 1})^2}{a_{j_2j_2 0}[(y_{j_1}-a_{j_2j_2 1})^2+b_{j_2j_2}]}\Big|_{y^*_{j_1}=a_{j_1j_1 1}} = 0,$$
$$\frac{d^2x_{j_1}}{dx^2_{j_2}}\Big|_{y^*_{j_1}=a_{j_1j_1 1}} = 0.$$

Thus, if

$$\frac{d^3x_{j_1}}{dx^3_{j_2}}\Big|_{y^*_{j_1}=a_{j_1j_1 1}} = \frac{a_{j_1j_1 0}\alpha_{j_1j_2}\alpha_{j_1j_2}}{a_{j_2j_2 0}[(a_{j_1j_1 1}-a_{j_2j_2 1})^2+b_{j_2j_2}]} > 0,$$

there is an increasing-inflection flow at $y^*_{j_1}=a_{j_1j_1 1}$; and if

$$\frac{d^3x_{j_1}}{dx^3_{j_2}}\Big|_{y^*_{j_1}=a_{j_1j_1 1}} = \frac{a_{j_1j_1 0}\alpha_{j_1j_2}\alpha_{j_1j_2}}{a_{j_2j_2 0}[(a_{j_1j_1 1}-a_{j_2j_2 1})^2+b_{j_2j_2}]} < 0,$$

there is a decreasing-inflection flow at $y^*_{j_1}=a_{j_1j_1 1}$. Let

$$\dot{x}^*_{j_2} = a_{j_2j_2 0}[(a_{j_1j_1 1}-a_{j_2j_2 1})^2+b_{j_2j_2}].$$

The inflection flows are positive and negative for $\dot{x}^*_{j_2}>0$ and $\dot{x}^*_{j_2}<0$, respectively.

Therefore, the flows at $y^*_{j_1}=a_{j_1j_1 1}$ in the x_{j_1} -direction and the x_{j_2}-direction have the following properties as in Eqs. (4.14)–(4.17).

- For $a_{j_1j_1 0}\alpha^2_{j_1j_2}>0$ and $a_{j_2j_2 0}>0$,

$$(a_{j_1j_1 1},\dot{x}^*_{j_2}) = \underbrace{(\text{II},\text{pF})}_{\text{increasing-inflection flow }(+)}.$$

- For $a_{j_1j_1 0}\alpha^2_{j_1j_2}<0$ and $a_{j_2j_2 0}>0$,

$$(a_{j_1j_1 1},\dot{x}^*_{j_2}) = \underbrace{(\text{DI},\text{pF})}_{\text{decreasing-inflection flow }(+)}.$$

- For $a_{j_1j_1 0}\alpha^2_{j_1j_2}>0$ and $a_{j_2j_2 0}>0$,

$$(a_{j_1j_1 1},\dot{x}^*_{j_2}) = \underbrace{(\text{DI},\text{nF})}_{\text{decreasing-inflection flow }(-)}.$$

- For $a_{j_1j_10}\alpha^2_{j_1j_2} < 0$ and $a_{j_2j_20} > 0$,

$$(a_{j_1j_11}, \dot{x}^*_{j_2}) = \underbrace{(\mathrm{II}, \mathrm{nF})}_{\text{increasing-inflection flow }(-)}.$$

(iii) Consider a dynamical system as

$$\begin{aligned}\dot{x}_{j_1} &= a_{j_1j_10}(y_{j_1} - a_{j_1j_11})(y_{j_1} - a_{j_1j_12}),\\ \dot{x}_{j_2} &= a_{j_2j_20}[(y_{j_1} - a_{j_2j_21})^2 + b_{j_2j_2}].\end{aligned}$$

The dynamical systems in the y_{j_1}-direction and the $y^{\perp}_{j_1}$-direction can be easily constructed as Eq. (4.21).

From the differential equations of the dynamical systems, at $y^*_{j_1} = a_{j_1j_1s} \neq a_{j_1j_1l}$,

$$\frac{dx_{j_1}}{dx_{j_2}}\bigg|_{y^*_{j_1}=a_{j_1j_1s}} = \frac{a_{j_1j_10}(y_{j_1} - a_{j_1j_1s})(y_{j_1} - a_{j_1j_1l})}{a_{j_2j_20}[(y_{j_1} - a_{j_2j_21})^2 + b_{j_2j_2}]}\bigg|_{y^*_{j_1}=a_{j_1j_1s}} = 0.$$

If

$$\frac{d^2x_{j_1}}{dx^2_{j_2}}\bigg|_{y^*_{j_1}=a_{j_1j_1s}} = \frac{a_{j_1j_10}\alpha_{j_1j_2}(a_{j_1j_1s} - a_{j_1j_1l})}{a_{j_2j_20}[(a_{j_1j_1s} - a_{j_2j_21})^2 + b_{j_2j_2}]} > 0,$$

there is an up-parabola flow at $y^*_{j_1} = a_{j_1j_1s}$ in the x_{j_1}-direction. If

$$\frac{d^2x_{j_1}}{dx^2_{j_2}}\bigg|_{y^*_{j_1}=a_{j_1j_1s}} = \frac{a_{j_1j_10}\alpha_{j_1j_2}(a_{j_1j_1s} - a_{j_1j_1l})}{a_{j_2j_20}[(a_{j_1j_1s} - a_{j_2j_21})^2 + b_{j_2j_2}]} < 0,$$

there is a down-parabola flow at $y^*_{j_1} = a_{j_1j_1s}$ in the x_{j_1}-direction. Let

$$\dot{x}^*_{j_2} = a_{j_2j_20}[(a_{j_1j_1s} - a_{j_2j_21})^2 + b_{j_2j_2}].$$

For $a_{j_2j_20} > 0$, $\dot{x}^*_{j_2} > 0$ implies the positive flow in the x_{j_2}-direction exists. For $a_{j_2j_20} < 0$, $\dot{x}^*_{j_2} < 0$ implies the negative flow in the x_{j_2}-direction exists.

Thus, the equilibrium of $y^*_{j_1} = a_{j_1j_1s}$ $(s,\ l \in \{1,2\},\ s \neq l)$ in the x_{j_1}-direction possesses the one-dimensional flow properties in Eqs. (4.23)–(4.26).

- For $a_{j_1j_10}\alpha_{j_1j_2}(a_{j_1j_1s} - a_{j_1j_1l}) > 0$ and $a_{j_2j_20} > 0$,

$$(a_{j_1j_1s}, \dot{x}^*_{j_2}) = \underbrace{(\mathrm{UP}, \mathrm{pF})}_{\text{up-parabola flow }(+)}.$$

- For $a_{j_1j_10}\alpha_{j_1j_2}(a_{j_1j_1s} - a_{j_1j_1l}) < 0$ and $a_{j_2j_20} > 0$,

$$(a_{j_1j_1s}, \dot{x}^*_{j_2}) = \underbrace{(\text{DP}, \text{pF})}_{\text{down-parabola flow } (+)}.$$

- For $a_{j_1j_10}\alpha_{j_1j_2}(a_{j_1j_1s} - a_{j_1j_1l}) > 0$ and $a_{j_2j_20} < 0$,

$$(a_{j_1j_1s}, \dot{x}^*_{j_2}) = \underbrace{(\text{DP}, \text{nF})}_{\text{down-parabola flow } (-)}.$$

- For $a_{j_1j_10}\alpha_{j_1j_2}(a_{j_1j_1s} - a_{j_1j_1l}) < 0$ and $a_{j_2j_20} < 0$,

$$(a_{j_1j_1s}, \dot{x}^*_{j_2}) = \underbrace{(\text{UP}, \text{nF})}_{\text{up-parabola flow } (-)}.$$

(iv) From a dynamical system of

$$\begin{aligned} \dot{x}_{j_1} &= a_{j_1j_10}(y_{j_1} - a_{j_1j_11})(y_{j_1} - a_{j_1j_12}), \\ \dot{x}_{j_2} &= a_{j_2j_20}(y_{j_1} - a_{j_2j_21})^2. \end{aligned}$$

The corresponding dynamical systems in the y_{j_1}-direction and the $y^{\perp}_{j_1}$-direction are obtained in Eqs. (4.30) and (4.31).

(iv$_1$) From the differential equation, for $y^*_{j_1} = a_{j_1j_1s}(s, l \in \{1, 2\}, s \neq l)$,

$$\frac{dx_{j_1}}{dx_{j_2}}\bigg|_{y^*_{j_1} = a_{j_1j_1s}} = \frac{a_{j_1j_10}(y_{j_1} - a_{j_1j_1s})(y_{j_1} - a_{j_1j_1l})}{a_{j_2j_20}(y_{j_1} - a_{j_2j_21})^2}\bigg|_{y^*_{j_1} = a_{j_1j_1s}} = 0.$$

If

$$\frac{d^2x_{j_1}}{dx^2_{j_2}}\bigg|_{y^*_{j_1} = a_{j_1j_1s}} = \frac{a_{j_1j_10}\alpha_{j_1j_2}(a_{j_1j_1s} - a_{j_1j_1l})}{a_{j_2j_20}(a_{j_1j_1s} - a_{j_2j_21})^2} > 0,$$

there is an up-parabola flow at $y^*_{j_1} = a_{j_1j_1s}$ in the x_{j_1}-direction. If

$$\frac{d^2x_{j_1}}{dx^2_{j_2}}\bigg|_{y^*_{j_1} = a_{j_1j_1s}} = \frac{a_{j_1j_10}\alpha_{j_1j_2}(a_{j_1j_1s} - a_{j_1j_1l})}{a_{j_2j_20}(a_{j_1j_1s} - a_{j_2j_21})^2} < 0,$$

there is a down-parabola flow at $y^*_{j_1} = a_{j_1j_1s}$ in the x_{j_1}-direction. Let

$$\dot{x}_{j_2}^* = a_{j_2j_20}(a_{j_1j_1s} - a_{j_2j_21})^2.$$

For $a_{j_2j_20} > 0$, $\dot{x}_{j_2}^* > 0$ implies that the positive flow in the x_{j_2}-direction exists. For $a_{j_2j_20} < 0$, $\dot{x}_{j_2}^* < 0$ implies that the negative flow in the x_{j_2}-direction exists.

Therefore, the simple equilibriums of $y_{j_1}^* = a_{j_1j_1s}(s,\ l \in \{1,2\},\ s \neq l)$ for the x_{j_1}-direction have four cases with $\dot{x}_{j_2}^* = a_{220}(a_{j_1j_1s} - a_{j_2j_21})^2$ as in Eqs. (4.32)–(4.35).

- For $a_{j_1j_10}\alpha_{j_1j_2}(a_{j_1j_1s} - a_{j_1j_1l}) > 0$ and $a_{j_2j_20}(a_{j_1j_1s} - a_{j_2j_21})^2 > 0$,

$$(a_{j_1j_1s}, \dot{x}_{j_2}^*) = \underbrace{(\text{UP}, \text{pF})}_{\text{up-parabola flow } (+)}.$$

- For $a_{j_1j_10}\alpha_{j_1j_2}(a_{j_1j_1s} - a_{j_1j_1l}) < 0$ and $a_{j_2j_20}(a_{j_1j_1s} - a_{j_2j_21})^2 > 0$,

$$(a_{j_1j_1s}, \dot{x}_{j_2}^*) = \underbrace{(\text{DP}, \text{pF})}_{\text{Down-parabola flow } (+)}.$$

- For $a_{j_1j_10}\alpha_{j_1j_2}(a_{j_1j_1s} - a_{j_1j_1l}) > 0$ and $a_{j_2j_20}(a_{j_1j_1s} - a_{j_2j_21})^2 < 0$,

$$(a_{j_1j_1s}, \dot{x}_{j_2}^*) = \underbrace{(\text{DP}, \text{nF})}_{\text{Down-parabola flow } (-)}.$$

- For $a_{j_1j_10}\alpha_{j_1j_2}(a_{j_1j_1s} - a_{j_1j_1l}) < 0$ and $a_{j_2j_20}(a_{j_1j_1s} - a_{j_2j_21})^2 < 0$,

$$(a_{j_1j_1s}, \dot{x}_{j_2}^*) = \underbrace{(\text{UP}, \text{nF})}_{\text{up-parabola flow } (-)}.$$

(iv$_2$) For $y_{j_1}^* = a_{j_2j_21}$,

$$\begin{aligned}
&\frac{dx_{j_2}}{dx_{j_1}}\bigg|_{y_{j_1}^* = a_{j_2j_21}} = \frac{a_{j_2j_20}(y_{j_1} - a_{j_2j_21})^2}{a_{j_1j_10}(y_{j_1} - a_{j_1j_1s})(y_{j_1} - a_{j_1j_1l})}\bigg|_{y_{j_1}^* = a_{j_2j_21}} = 0,\\
&\frac{d^2x_{j_2}}{dx_{j_1}^2}\bigg|_{y_{j_1}^* = a_{j_2j_21}} = 0.
\end{aligned}$$

If

$$\frac{d^3x_{j_2}}{dx_{j_1}^3}\bigg|_{y_{j_1}^* = a_{j_1j_1s}} = \frac{2a_{j_2j_20}\alpha_{j_1j_2}\alpha_{j_1j_2}}{a_{j_1j_10}(a_{j_2j_21} - a_{j_1j_11})(a_{j_2j_21} - a_{j_1j_12})} > 0,$$

there is an increasing-inflection flow at $y_{j_1}^* = a_{j_2j_21}$. If

$$\frac{d^3 x_{j_2}}{dx_{j_1}^3}\Big|_{y_{j_1}^*=a_{j_1j_1s}} = \frac{2a_{j_2j_20}\alpha_{j_1j_2}\alpha_{j_1j_2}}{a_{j_1j_10}(a_{j_2j_21}-a_{j_1j_11})(a_{j_2j_21}-a_{j_1j_12})} < 0,$$

there is a decreasing-inflection flow at $y_{j_1}^* = a_{j_2j_21}$. Let

$$\dot{x}_{j_1}^* = a_{j_1j_10}\prod\nolimits_{s=1}^{2}(a_{j_2j_21}-a_{j_1j_1s}).$$

For $\dot{x}_{j_1}^* < 0$ and $\dot{x}_{j_1}^* < 0$, the inflection flows are positive and negative in the x_{j_1}-direction.

Thus, the equilibriums of $y_{j_1}^* = a_{j_2j_21}$ for the x_{j_2}-direction have the following properties as in Eqs. (4.36)–(4.39).

- For $a_{j_1j_10}\prod_{s=1}^{2}(a_{j_2j_21}-a_{j_1j_1s}) > 0$ and $a_{j_2j_20}\alpha_{j_1j_2}^2 > 0$,

$$(\dot{x}_{j_1}^*, a_{j_2j_21}) = \underbrace{(\text{pF}, \text{II})}_{\text{increaseing-inflection flow }(+)}.$$

- For $a_{j_1j_10}\prod_{s=1}^{2}(a_{j_2j_21}-a_{j_1j_1s}) < 0$ and $a_{j_2j_20}\alpha_{j_1j_2}^2 > 0$,

$$(\dot{x}_{j_1}^*, a_{j_2j_21}) = \underbrace{(\text{nF}, \text{DI})}_{\text{decreaseing-inflection flow }(-)}.$$

- For $a_{j_1j_10}\prod_{s=1}^{2}(a_{j_2j_21}-a_{j_1j_1s}) > 0$ and $a_{j_2j_20}\alpha_{j_1j_2}^2 < 0$,

$$(\dot{x}_{j_1}^*, a_{j_2j_21}) = \underbrace{(\text{pF}, \text{DI})}_{\text{decreaseing-inflection flow }(+)}.$$

- For $a_{j_1j_10}\prod_{s=1}^{2}(a_{j_2j_21}-a_{j_1j_1s}) < 0$ and $a_{j_2j_20}\alpha_{j_1j_2}^2 < 0$,

$$(\dot{x}_{j_1}^*, a_{j_2j_21}) = \underbrace{(\text{nF}, \text{II})}_{\text{increaseing-inflection flow }(-)}.$$

(iv$_3$) For $a_{j_2j_21} = a_{j_1j_1s}$ ($s, l \in \{1,2\}$, $s \neq l$), consider

$$\begin{aligned}\dot{x}_{j_1} &= a_{j_1j_10}(y_{j_1}-a_{j_1j_1s})(y_{j_1}-a_{j_1j_1l}),\\ \dot{x}_{j_2} &= a_{j_2j_20}(y_{j_1}-a_{j_2j_21})^2.\end{aligned}$$

The corresponding dynamical systems in the y_{j_1}-direction and in the $y_{j_1}^{\perp}$-direction are obtained as in Eqs. (4.41) and (4.42).

For $y_{j_1}^* = a_{j_2j_21} = a_{j_1j_1s}$, the flow at $y_{j_1}^* = a_{j_1j_1s}$ is considered through

$$\frac{dx_{j_2}}{dx_{j_1}}\Big|_{y_{j_1}^*=a_{j_1j_1s}}=\frac{a_{j_2j_20}(y_{j_1}-a_{j_2j_2s})}{a_{j_1j_10}(y_{j_1}-a_{j_1j_1l})}\Big|_{y_{j_1}^*=a_{j_1j_1s}}=0.$$

If

$$\frac{d^2x_{j_2}}{dx_{j_1}^2}\Big|_{y_{j_1}^*=a_{j_1j_1s}}=\frac{a_{j_2j_20}\alpha_{j_1j_1}}{a_{j_1j_10}(a_{j_1j_1s}-a_{j_1j_1l})}=\frac{a_{j_2j_20}\alpha_{j_1j_1}\alpha_{j_1j_1}}{\alpha_{j_1j_1}a_{j_1j_10}(a_{j_1j_1s}-a_{j_1j_1l})}>0,$$

there is an up-parabola flow at $y_{j_1}^*=a_{j_1j_1s}$. If

$$\frac{d^2x_{j_2}}{dx_{j_1}^2}\Big|_{y_{j_1}^*=a_{j_1j_1s}}=\frac{a_{j_2j_20}\alpha_{j_1j_1}}{a_{j_1j_10}(a_{j_1j_1s}-a_{j_1j_1l})}=\frac{a_{j_2j_20}\alpha_{j_1j_1}\alpha_{j_1j_1}}{\alpha_{j_1j_1}a_{j_1j_10}(a_{j_1j_1s}-a_{j_1j_1l})}<0,$$

there is a down-parabola flow at $y_{j_1}^*=a_{j_1j_1s}$. In the vicinity of $y_{j_1}^*=a_{j_1j_1s}=a_{j_2j_21}$, we have

$$\Delta\dot{y}_{j_1}=\alpha_{j_1j_1}a_{j_1j_10}(a_{j_1j_1s}-a_{j_1j_1l})\Delta y_{j_1}=G_{j_1j_10}\Delta y_{j_1},$$

where

$$G_{j_1j_10}=\alpha_{j_1j_1}a_{j_1j_10}(a_{j_1j_1s}-a_{j_1j_1l}).$$

The parabola flows of $y_{j_1}^*=a_{j_1j_1s}=a_{j_2j_21}$ are source and sink for $G_{j_1j_10}>0$ and $G_{j_1j_10}<0$, respectively.

Therefore, the infinite-equilibriums of $y_{j_1}^*=a_{j_1j_1s}=a_{j_2j_21}$ for the x_{j_1}-direction have the following properties as in Eqs. (4.44)–(4.47).

- For $\alpha_{j_1j_1}a_{j_1j_10}(a_{j_1j_1s}-a_{j_2j_2l})>0$ and $a_{j_2j_20}\alpha_{j_1j_1}^2>0$,

$$(a_{j_1j_1s},a_{j_2j_21})=\underbrace{(\text{SO},\text{UP})}_{\text{infinite-equilibrium}}.$$

- For $\alpha_{j_1j_1}a_{j_1j_10}(a_{j_1j_1s}-a_{j_2j_2l})<0$ and $a_{j_2j_20}\alpha_{j_1j_1}^2>0$,

$$(a_{j_1j_1s},a_{j_2j_21})=\underbrace{(\text{SI},\text{DP})}_{\text{infinite-equilibrium}}.$$

- For $\alpha_{j_1j_1}a_{j_1j_10}(a_{j_1j_1s}-a_{j_2j_2l})>0$ and $a_{j_2j_20}\alpha_{j_1j_1}^2<0$,

$$(a_{j_1j_1s},a_{j_2j_21})=\underbrace{(\text{SO},\text{DP})}_{\text{infinite-equilibrium}}.$$

- For $\alpha_{j_1j_1}a_{j_1j_10}(a_{j_1j_1s}-a_{j_2j_2l})<0$ and $a_{j_2j_20}\alpha^2_{j_1j_1}<0$,

$$(a_{j_1j_1s},a_{j_2j_21})=\underbrace{(\text{SI, UP})}_{\text{infinite-equilibrium}}.$$

(v) Consider

$$\begin{aligned}&\dot{x}_{j_1}=a_{j_1j_10}(y_{j_1}-a_{j_1j_11})(y_{j_1}-a_{j_1j_12}),\\&\dot{x}_{j_2}=a_{j_2j_20}(y_{j_1}-a_{j_2j_21})(y_{j_1}-a_{j_2j_22}),\\&j_1,j_2\in\{1,2\};j_1\neq j_2.\end{aligned}$$

The corresponding dynamical systems in the y_{j_1}-direction and in the $y_{j_1}^{\perp}$-direction are obtained straightforward in Eqs. (4.51) and (4.52)
(v_1) For $y_{j_1}^*=a_{j_1j_1s_1}$(s_1, $s_2\in\{1,2\}$, $s_2\neq s_1$), the following relation exists.

$$\frac{dx_{j_1}}{dx_{j_2}}\bigg|_{y_{j_1}^*=a_{j_1j_1s_1}}=\frac{a_{j_1j_10}(y_{j_1}-a_{j_1j_1s_1})(y_{j_1}-a_{j_1j_1s_2})}{a_{j_2j_20}(y_{j_1}-a_{j_2j_2l_1})(y_{j_1}-a_{j_2j_2l_2})}\bigg|_{y_{j_1}^*=a_{j_1j_1s_1}}=0.$$

If

$$\frac{d^2x_{j_1}}{dx_{j_2}^2}\bigg|_{y_{j_1}^*=a_{j_1j_1s_1}}=\frac{\alpha_{j_1j_2}a_{j_1j_10}(a_{j_1j_1s_1}-a_{j_1j_1s_2})}{a_{j_2j_20}(a_{j_1j_1s_1}-a_{j_2j_2l_1})(a_{j_1j_1s_1}-a_{j_2j_2l_2})}>0,$$

there is an up-parabola flow at $y_{j_1}^*=a_{j_1j_1s_1}$ for the x_{j_1}-direction. If

$$\frac{d^2x_{j_1}}{dx_{j_2}^2}\bigg|_{y_{j_1}^*=a_{j_1j_1s_1}}=\frac{\alpha_{j_1j_2}a_{j_1j_10}(a_{j_1j_1s_1}-a_{j_1j_1s_2})}{a_{j_2j_20}(a_{j_1j_1s_1}-a_{j_2j_2l_1})(a_{j_1j_1s_1}-a_{j_2j_2l_2})}<0,$$

there is a down-parabola flow at $y_{j_1}^*=a_{j_1j_1s_1}$ for the x_{j_1}-direction. Let

$$\dot{x}_{j_2}^*=a_{j_2j_20}(a_{j_1j_1s_1}-a_{j_2j_2l_1})(a_{j_1j_1s_1}-a_{j_2j_2l_2})$$

For $\dot{x}_{j_2}^*>0$ and $\dot{x}_{j_2}^*<0$, the flows at $y_{j_1}^*=a_{j_1j_1s_1}$ for the x_{j_2}-direction are positive and negative, respectively.

Therefore, the equilibrium of $y_{j_1}^*=a_{j_1j_1s_1},a_{j_1j_1s_2}$ (s_1, $s_2\in\{1,2\}$, $s_2\neq s_1$) for the x_{j_1}-direction has four cases with $\dot{x}_{j_2}^*=a_{j_2j_20}\prod_{l_1=1}^{2}(a_{j_1j_1s_1}-a_{j_2j_2l_1})$ as presented in Eqs. (4.53)–(4.56).

- For $\alpha_{j_1j_2}a_{j_1j_10}(a_{j_1j_2s_1}-a_{j_1j_2s_2})>0$ and $a_{j_2j_20}\prod_{l_1=1}^{2}(a_{j_1j_1s_1}-a_{j_2j_2l_1})>0$,

$$(a_{j_1j_1s_1},\dot{x}_{j_2}^{*})=\underbrace{(\text{UP},\text{pF})}_{\text{up-parabola flow }(+)}.$$

- For $\alpha_{j_1j_2}a_{j_1j_10}(a_{j_1j_2s_1}-a_{j_1j_2s_2})<0$ and $a_{j_2j_20}\prod_{l_1=1}^{2}(a_{j_1j_1s_1}-a_{j_2j_2l_1})>0$,

$$(a_{j_1j_1s_1},\dot{x}_{j_2}^{*})=\underbrace{(\text{DP},\text{pF})}_{\text{down-parabola flow }(+)}.$$

- For $\alpha_{j_1j_2}a_{j_1j_10}(a_{j_1j_2s_1}-a_{j_1j_2s_2})>0$ and $a_{j_2j_20}\prod_{l_1=1}^{2}(a_{j_1j_1s_1}-a_{j_2j_2l_1})<0$,

$$(a_{j_1j_1s_1},\dot{x}_{j_2}^{*})=\underbrace{(\text{DP},\text{nF})}_{\text{down-parabola flow }(-)}.$$

- For $\alpha_{j_1j_2}a_{j_1j_10}(a_{j_1j_2s_1}-a_{j_1j_2s_2})<0$ and $a_{j_2j_20}\prod_{l_1=1}^{2}(a_{j_1j_1s_1}-a_{j_2j_2l_1})<0$,

$$(a_{j_1j_1s_1},\dot{x}_{j_2}^{*})=\underbrace{(\text{UP},\text{nF})}_{\text{up-parabola flow }(-)}.$$

(v_2) Similar to (vi_1), for $y_{j_1}^{*}=a_{j_2j_2l_1}$ (l_1, $l_2\in\{1,2\}$, $l_1\neq l_2$), the following relation exists.

$$\frac{dx_{j_2}}{dx_{j_1}}\Big|_{y_{j_1}^{*}=a_{j_2j_2l_1}}=\frac{a_{j_2j_20}(y_{j_1}-a_{j_2j_2l_1})(y_{j_1}-a_{j_2j_2l_2})}{a_{j_1j_10}(y_{j_1}-a_{j_1j_1s_1})(y_{j_1}-a_{j_1j_1s_2})}\Big|_{y_{j_1}^{*}=a_{j_2j_2l_1}}=0.$$

If

$$\frac{d^2x_{j_2}}{dx_{j_1}^2}\Big|_{y_{j_1}^{*}=a_{j_2j_2l_1}}=\frac{\alpha_{j_1j_1}a_{j_2j_20}(a_{j_2j_2l_1}-a_{j_2j_2l_2})}{a_{j_1j_10}(a_{j_2j_2l_1}-a_{j_1j_1s_1})(a_{j_2j_2l_1}-a_{j_1j_1s_2})}>0,$$

there is an up-parabola flow at $y_{j_1}^{*}=a_{j_2j_2l_1}$ for the x_{j_2}-direction. If

$$\frac{d^2x_{j_2}}{dx_{j_1}^2}\Big|_{y_{j_1}^{*}=a_{j_2j_2l_1}}=\frac{\alpha_{j_1j_1}a_{j_2j_20}(a_{j_2j_2l_1}-a_{j_2j_2l_2})}{a_{j_1j_10}(a_{j_2j_2l_1}-a_{j_1j_1s_1})(a_{j_2j_2l_1}-a_{j_1j_1s_2})}<0,$$

there is a down-parabola flow at $y_{j_1}^{*}=a_{j_2j_2l_1}$ for the x_{j_2}-direction. Let

$$\dot{x}_{j_1}^{*}=a_{j_1j_10}\prod_{s_1=1}^{2}(a_{j_2j_2l_1}-a_{j_1j_1s_1}).$$

For $\dot{x}_{j_2}^{*}>0$ and $\dot{x}_{j_2}^{*}<0$, the flows at $y_{j_1}^{*}=a_{j_1j_1s_1}$ for the x_{j_2}-direction are positive and negative, respectively.

Thus, the equilibrium of $y_{j_1}^* = a_{j_2j_2l_1}$ ($l_1, l_2 \in \{1,2\}, l_1 \neq l_2$) for the x_{j_2}-direction has four cases with $\dot{x}_{j_1}^* = a_{j_1j_10}\prod_{s_1=1}^{2}(a_{j_2j_2l_1} - a_{j_1j_1s_1})$, as presented in Eqs. (4.57)–(4.60).

- For $a_{j_1j_10}\prod_{s_1=1}^{2}(a_{j_2j_2l_1} - a_{j_1j_1s_1}) > 0$ and $\alpha_{j_1j_1}a_{j_2j_20}(a_{j_2j_2l_1} - a_{j_2j_2l_2}) > 0$,

$$(\dot{x}_{j_1}^*, a_{j_2j_2l_1}) = \underbrace{(\mathrm{pF}, \mathrm{UP})}_{\text{up-parabola flow }(+)}.$$

- For $a_{j_1j_10}\prod_{s_1=1}^{2}(a_{j_2j_2l_1} - a_{j_1j_1s_1}) < 0$ and $\alpha_{j_1j_1}a_{j_2j_20}(a_{j_2j_2l_1} - a_{j_2j_2l_2}) > 0$,

$$(\dot{x}_{j_1}^*, a_{j_2j_2l_1}) = \underbrace{(\mathrm{nF}, \mathrm{DP})}_{\text{down-parabola flow }(-)}.$$

- For $a_{j_1j_10}\prod_{s_1=1}^{2}(a_{j_2j_2l_1} - a_{j_1j_1s_1}) > 0$ and $\alpha_{j_1j_1}a_{j_2j_20}(a_{j_2j_2l_1} - a_{j_2j_2l_2}) < 0$,

$$(\dot{x}_{j_1}^*, a_{j_2j_2l_1}) = \underbrace{(\mathrm{pF}, \mathrm{DP})}_{\text{down-parabola flow }(+)}.$$

- For $a_{j_1j_10}\prod_{s_1=1}^{2}(a_{j_2j_2l_1} - a_{j_1j_1s_1}) < 0$ and $\alpha_{j_1j_1}a_{j_2j_20}(a_{j_2j_2l_1} - a_{j_2j_2l_2}) < 0$,

$$(\dot{x}_{j_1}^*, a_{j_2j_2l_1}) = \underbrace{(\mathrm{nF}, \mathrm{UP})}_{\text{up-parabola flow }(-)}.$$

(v_3) For $a_{j_1j_1s_1} = a_{j_1j_1l_1}$($s_1$, $l_1 \in \{1,2\}$), consider

$$\begin{aligned}
\dot{x}_{j_1} &= a_{j_1j_10}(y_{j_1} - a_{j_1js_1})(y_{j_1} - a_{j_1js_2}),\\
\dot{x}_{j_2} &= a_{j_2j_20}(y_{j_1} - a_{j_2j_2l_1})(y_{j_1} - a_{j_2j_2l_2}).
\end{aligned}$$

Similarly, the corresponding dynamical systems in the y_{j_1}-direction and in the $y_{j_1}^{\perp}$-direction are straightforward obtained in Eqs. (4.62) and (4.63).

For $y_{j_1}^* = a_{j_1j_1s_1} = a_{j_2j_2l_1}$ with $a_{j_1j_1s_2} \neq a_{j_2j_2l_2}$, if

$$\left.\frac{dx_{j_2}}{dx_{j_1}}\right|_{y_{j_1}^* = a_{j_1j_1s_1}} = \frac{a_{j_2j_20}(a_{j_2j_2l_1} - a_{j_2j_2l_2})}{a_{j_1j_10}(a_{j_1j_1s_1} - a_{j_1j_1s_2})} > 0,$$

there is an increasing-inflection flow at $y_{j_1}^* = a_{j_1j_1s_1} = a_{j_2j_2l_1}$, and if

$$\left.\frac{dx_{j_2}}{dx_{j_1}}\right|_{y_{j_1}^* = a_{j_1j_1s_1}} = \frac{a_{j_2j_20}(a_{j_2j_2l_1} - a_{j_2j_2l_2})}{a_{j_1j_10}(a_{j_1j_1s_1} - a_{j_1j_1s_2})} < 0,$$

there is a decreasing-inflection flow at $y_{j_1}^* = a_{j_1j_1s_1}$.

In the vicinity of $y^*_{j_1} = a_{j_1j_1s_1} = a_{j_2j_2l_1}$, the variational equation is

$$\begin{aligned}\Delta\dot{y}_{j_1} &= [\alpha_{j_1j_1}a_{j_1j_10}(a_{j_1j_1s_1} - a_{j_1j_1s_2}) + \alpha_{j_1j_2}a_{j_2j_20}(a_{j_1j_1l_1} - a_{j_1j_1l_2})]\Delta y_{j_1} \\ &= G_{j_1j_10}\Delta y_{j_1},\end{aligned}$$

where

$$G_{j_1j_10} = \alpha_{j_1j_1}a_{j_1j_10}(a_{j_1j_1s_1} - a_{j_1j_1s_2}) + \alpha_{j_1j_2}a_{j_2j_20}(a_{j_1j_1l_1} - a_{j_1j_1l_2}).$$

The inflection flows at $y^*_{j_1} = a_{j_1j_1s_1}$ are source and sink for $G_{j_1j_10} > 0$ and $G_{j_1j_10} < 0$, respectively.

Therefore, the infinite-equilibrium of $y^*_{j_1} = a_{j_1j_1s_1} = a_{j_2j_2l_1}$ with $a_{j_1j_1s_2} \neq a_{j_2j_1l_2}$ has four cases as given by Eqs. (4.65)–(4.68).

- For $a_{j_1j_10}(a_{j_1j_1s_1} - a_{j_1j_1s_2})/a_{j_2j_20}(a_{j_2j_2l_1} - a_{j_2j_2l_2}) > 0$ and $G_{j_1j_10} > 0$,

$$(a_{j_1j_1s_1}, a_{j_2j_2l_1}) = \underbrace{(\text{II}, \text{SO})}_{\text{increasing-inflection source}}.$$

- For $a_{j_1j_10}(a_{j_1j_1s_1} - a_{j_1j_1s_2})/a_{j_2j_20}(a_{j_2j_2l_1} - a_{j_2j_2l_2}) > 0$ and $G_{j_1j_10} < 0$,

$$(a_{j_1j_1s_1}, a_{j_2j_2l_1}) = \underbrace{(\text{II}, \text{SI})}_{\text{increasing-inflection sink}}.$$

- For $a_{j_1j_10}(a_{j_1j_1s_1} - a_{j_1j_1s_2})/a_{j_2j_20}(a_{j_2j_2l_1} - a_{j_2j_2l_2}) < 0$ and $G_{j_1j_10} > 0$,

$$(a_{j_1j_1s_1}, a_{j_2j_2l_1}) = \underbrace{(\text{DI}, \text{SO})}_{\text{decreasing-inflection source}}.$$

- For $a_{j_1j_10}(a_{j_1j_1s_1} - a_{j_1j_1s_2})/a_{j_2j_20}(a_{j_2j_2l_1} - a_{j_2j_2l_2}) < 0$ and $G_{j_1j_10} < 0$,

$$(a_{j_1j_1s_1}, a_{j_2j_2l_1}) = \underbrace{(\text{DI}, \text{SI})}_{\text{decreasing-inflection sink}}.$$

The infinite-equilibriums are the switching bifurcations for the down-parabola and up-parabola flows with the up-parabola and down-parabola flows.

(v_4) For $a_{j_1j_1s_1} = a_{j_2j_2l_1}$ and $a_{j_1j_1s_2} = a_{j_2j_2l_2}$ (s_1, s_2, l_1, $l_2 \in \{1, 2\}$, $s_1 \neq s_2$, $l_1 \neq l_2$) with $a_{j_1j_1s_1} \neq a_{j_1j_1s_2}$ consider

$$\begin{aligned}\dot{x}_{j_1} &= a_{j_1j_10}(y_{j_1} - a_{j_1j_1s_1})(y_{j_1} - a_{j_1j_1s_2}), \\ \dot{x}_{j_2} &= a_{j_2j_20}(y_{j_1} - a_{j_2j_2l_1})(y_{j_1} - a_{j_2j_2l_2}).\end{aligned}$$

The corresponding dynamical systems in the y_{j_1}-direction and in the $y_{j_1}^{\perp}$-direction are easily constructed as in Eqs. (4.70) and (4.71)

For $a_{j_1j_1s_1} = a_{j_2j_2l_1}$ and $a_{j_1j_1s_2} = a_{j_2j_2l_2}$ if

$$\frac{dx_{j_2}}{dx_{j_1}}\bigg|_{y_{j_1}^* = a_{j_1j_1s_1}} = \frac{a_{j_2j_20}}{a_{j_1j_10}} > 0,$$

there is an increasing-inflection flow at $y_{j_1}^* = a_{j_1j_1s_1}$, and if

$$\frac{dx_{j_2}}{dx_{j_1}}\bigg|_{y_{j_1}^* = a_{j_1j_1s_1}} = \frac{a_{j_2j_20}}{a_{j_1j_10}} < 0,$$

there is a decreasing-inflection flow at $y_{j_1}^* = a_{j_1j_1s_1}$.

In the vicinity of $y_{j_1}^* = a_{j_1j_1s_1}$, we have

$$\Delta\dot{y}_{j_1} = \left[\alpha_{j_1j_1}a_{j_1j_10} + \alpha_{j_1j_2}a_{j_2j_20}\right](a_{j_1j_1s_1} - a_{j_1j_1s_2})\Delta y_{j_1} = G_{j_1j_10}\Delta y_{j_1},$$

where

$$G_{j_1j_10} = \left[\alpha_{j_1j_1}a_{j_1j_10} + \alpha_{j_1j_2}a_{j_2j_20}\right](a_{j_1j_1s_1} - a_{j_1j_1s_2}).$$

The inflection flows at $y_{j_1}^* = a_{j_1j_1s_1}$ are source and sink for $G_{j_1j_10} > 0$ and $G_{j_1j_10} < 0$, respectively.

Therefore, the infinite-equilibrium of $y_{j_1}^* = a_{j_1j_1s_1} = a_{j_2j_2l_1}$ and $a_{j_1j_1s_2} = a_{j_2j_2l_2}$ (s_1, s_2, $l_1, l_2 \in \{1,2\}$, $s_1 \neq s_2$, $l_1 \neq l_2$) with $a_{j_1j_1s_1} \neq a_{j_1j_1s_2}$ have four cases, which are obtained as presented in Eqs. (4.73)–(4.76).

- For $a_{j_1j_10}/a_{j_2j_20} > 0$ and $G_{j_1j_10} > 0$,

$$(a_{j_1j_1s_1}, a_{j_2j_2l_1}) = \underbrace{(\text{II}, \text{SO})}_{\text{increasing-inflection source}}.$$

- For $a_{j_1j_10}/a_{j_2j_20} > 0$ and $G_{j_1j_10} < 0$,

$$(a_{j_1j_1s_1}, a_{j_2j_2l_1}) = \underbrace{(\text{II}, \text{SI})}_{\text{increasing-inflection sink}}.$$

- For $a_{j_1j_10}/a_{j_2j_20} < 0$ and $G_{j_1j_10} > 0$,

$$(a_{j_1j_1s_1}, a_{j_2j_2l_1}) = \underbrace{(\text{DI}, \text{SO})}_{\text{decreasing-inflection source}}.$$

- For $a_{j_1 j_1 0}/a_{j_2 j_2 0} < 0$ and $G_{j_1 j_1 0} < 0$,

$$(a_{j_1 j_1 s_1}, a_{j_2 j_2 l_1}) = \underbrace{(\mathrm{DI}, \mathrm{SI})}_{\text{decreasing-inflection sink}}.$$

(vi) For $a_{j_1 j_1 1} \neq a_{j_2 j_2 1}$, consider

$$\begin{aligned} \dot{x}_{j_1} &= a_{j_1 j_1 0}(y_{j_1} - a_{j_1 j_1 1})^2, \\ \dot{x}_{j_2} &= a_{j_2 j_2 0}(y_{j_1} - a_{j_2 j_2 1})^2. \end{aligned}$$

Similarly, the corresponding dynamical systems in the y_{j_1}-direction and in the $y_{j_1}^{\perp}$-direction are straightforward constructed as in Eqs. (4.78) and (4.79).

(vi$_1$) For $y_{j_1}^* = a_{j_1 j_1 1}$, we have

$$\begin{aligned} \frac{dx_{j_1}}{dx_{j_2}}\bigg|_{y_{j_1}^* = a_{j_1 j_1 1}} &= \frac{a_{j_1 j_1 0}(y_{j_1} - a_{j_1 j_1 1})^2}{a_{j_2 j_2 0}(y_{j_1} - a_{j_2 j_2 1})^2}\bigg|_{y_{j_1}^* = a_{j_1 j_1 1}} = 0, \\ \frac{d^2 x_{j_1}}{dx_{j_2}^2}\bigg|_{y_{j_1}^* = a_{j_1 j_1 1}} &= 0. \end{aligned}$$

Thus, if

$$\frac{d^3 x_{j_1}}{dx_{j_2}^3}\bigg|_{y_{j_1}^* = a_{j_1 j_1 1}} = \frac{\alpha_{j_1 j_2} \alpha_{j_1 j_2} a_{j_1 j_1 0}}{a_{j_2 j_2 0}(a_{j_1 j_1 1} - a_{j_2 j_2 1})^2} > 0,$$

there is an increasing-inflection flow at $y_{j_1}^* = a_{j_1 j_1 1}$ in the x_{j_1}-direction, and if

$$\frac{d^3 x_{j_1}}{dx_{j_2}^3}\bigg|_{y_{j_1}^* = a_{j_1 j_1 1}} = \frac{\alpha_{j_1 j_2} \alpha_{j_1 j_2} a_{j_1 j_1 0}}{a_{j_2 j_2 0}(a_{j_1 j_1 1} - a_{j_2 j_2 1})^2} < 0,$$

there is a decreasing-inflection flow at $y_{j_1}^* = a_{j_1 j_1 1}$ in the x_{j_1}-direction. Let

$$\dot{x}_{j_2}^* = a_{j_2 j_2 0}(a_{j_1 j_1 1} - a_{j_2 j_2 1})^2.$$

The inflection flows at $y_{j_1}^* = a_{j_1 j_1 1}$ are positive and negative for $\dot{x}_{j_2}^* > 0$ and $\dot{x}_{j_2}^* < 0$, respectively.

Thus, the equilibrium of $y_{j_1}^* = a_{j_1 j_1 1}$ with $a_{j_1 j_1 1} \neq a_{j_2 j_2 1}$ have four cases, which are obtained as presented in Eqs. (4.80)–(4.83).

- For $a_{j_1j_10}>0$ and $a_{j_2j_20}(a_{j_1j_11}-a_{j_2j_21})^2>0$,

$$(a_{j_1j_11},\dot{x}_{j_2}^*)=\underbrace{(\mathrm{II},\mathrm{pF})}_{\text{increasing-inflection flow }(+)},$$

- For $a_{j_1j_10}<0$ and $a_{j_2j_20}(a_{j_1j_11}-a_{j_2j_21})^2>0$,

$$(a_{j_1j_11},\dot{x}_{j_2}^*)=\underbrace{(\mathrm{DI},\mathrm{pF})}_{\text{decreasing-inflection flow }(+)}.$$

- For $a_{j_1j_10}>0$ and $a_{j_2j_20}(a_{j_1j_11}-a_{j_2j_21})^2<0$,

$$(a_{j_1j_11},\dot{x}_{j_2}^*)=\underbrace{(\mathrm{DI},\mathrm{nF})}_{\text{decreasing-inflection flow }(-)}.$$

- For $a_{j_1j_10}<0$ and $a_{j_2j_20}(a_{j_1j_11}-a_{j_2j_21})^2<0$,

$$(a_{j_1j_11},\dot{x}_{j_2}^*)=\underbrace{(\mathrm{II},\mathrm{nF})}_{\text{increasing-inflection flow }(-)}.$$

(vi$_2$) For $y_{j_1}^*=a_{j_2j_21}$, we have

$$\frac{dx_{j_2}}{dx_{j_1}}\Big|_{y_{j_1}^*=a_{j_2j_21}}=\frac{a_{j_2j_20}(y_{j_1}-a_{j_2j_21})^2}{a_{j_1j_10}(y_{j_1}-a_{j_1j_11})^2}\Big|_{y_{j_1}^*=a_{j_2j_21}}=0,$$
$$\frac{d^2x_{j_2}}{dx_{j_1}^2}\Big|_{y_{j_1}^*=a_{j_2j_21}}=0.$$

Thus, if

$$\frac{d^3x_{j_2}}{dx_{j_1}^3}\Big|_{y_{j_1}^*=a_{j_2j_21}}=\frac{\alpha_{j_1j_1}\alpha_{j_1j_1}a_{j_2j_20}}{a_{j_1j_10}(a_{j_2j_21}-a_{j_1j_11})^2}>0,$$

there is an increasing-inflection flow at $y_{j_1}^*=a_{j_2j_21}$ in the x_{j_2}-direction, and if

$$\frac{d^3x_{j_2}}{dx_{j_1}^3}\Big|_{y_{j_1}^*=a_{j_2j_21}}=\frac{\alpha_{j_1j_1}\alpha_{j_1j_1}a_{j_2j_20}}{a_{j_1j_10}(a_{j_2j_21}-a_{j_1j_11})^2}<0.$$

there is a decreasing-inflection flow at $y_{j_1}^*=a_{j_2j_21}$ in the x_{j_2}-direction. Let

$$\dot{x}_{j_1}^*=a_{j_1j_10}(a_{j_2j_21}-a_{j_1j_11})^2.$$

The inflection flows at $y_{j_1}^*=a_{j_2j_21}$ are positive and negative for $\dot{x}_{j_1}^*>0$ and $\dot{x}_{j_1}^*<0$, respectively.

Thus, the equilibrium of $y_{j_1}^* = a_{j_2j_2 1}$ with $a_{j_1j_1 1} \neq a_{j_2j_2 1}$ have four cases, which are obtained as presented in Eqs. (4.84)–(4.87).

- For $a_{j_1j_1 0}(a_{j_2j_2 1} - a_{j_1j_1 1})^2 > 0$ and $a_{j_2j_2 0} > 0$,

$$(\dot{x}_{j_1}^*, a_{j_2j_2 1}) = \underbrace{(\mathrm{pF}, \mathrm{II})}_{\text{increasing-inflection flow } (+)}.$$

- For $a_{j_1j_1 0}(a_{j_2j_2 1} - a_{j_1j_1 1})^2 < 0$ and $a_{j_2j_2 0} > 0$,

$$(\dot{x}_{j_1}^*, a_{j_2j_2 1}) = \underbrace{(\mathrm{nF}, \mathrm{DI})}_{\text{decreasing-inflection flow } (-)}.$$

- For $a_{j_1j_1 0}(a_{j_2j_2 1} - a_{j_1j_1 1})^2 > 0$ and $a_{j_2j_2 0} < 0$,

$$(\dot{x}_{j_1}^*, a_{j_2j_2 1}) = \underbrace{(\mathrm{pF}, \mathrm{DI})}_{\text{decreasing-inflection flow } (+)}.$$

- For $a_{j_1j_1 0}(a_{j_2j_2 1} - a_{j_1j_1 1})^2 < 0$ and $a_{j_2j_2 0} < 0$,

$$(\dot{x}_{j_1}^*, a_{j_2j_2 1}) = \underbrace{(\mathrm{nF}, \mathrm{II})}_{\text{increasing-inflection flow } (-)}.$$

(vi$_3$) Consider a dynamical system with $a_{j_1j_1 1} = a_{j_2j_2 1}$,

$$\dot{x}_{j_1} = a_{j_1j_1 0}(y_{j_1} - a_{j_1j_1 1})^2,$$
$$\dot{x}_{j_2} = a_{j_2j_2 0}(y_{j_1} - a_{j_2j_2 1})^2.$$

Similarly, the corresponding dynamical systems in the y_{j_1}-direction and in the $y_{j_1}^{\perp}$-direction are straightforward constructed as in Eqs. (4.89) and (4.90).

For $y_{j_1}^* = a_{j_1j_1 1} = a_{j_2j_2 1}$, if

$$\left.\frac{dx_{j_2}}{dx_{j_1}}\right|_{y_{j_1}^* = a_{j_1j_1 1}} = \frac{a_{j_2j_2 0}}{a_{j_1j_1 0}} > 0,$$

there is an increasing-inflection flow at $y_{j_1}^* = a_{j_1j_1 1} = a_{j_2j_2 1}$, and if

$$\left.\frac{dx_{j_2}}{dx_{j_1}}\right|_{y_{j_1}^* = a_{j_1j_1 1}} = \frac{a_{j_2j_2 0}}{a_{j_1j_1 0}} < 0,$$

there is a decreasing-inflection flow at $y_{j_1}^* = a_{j_1j_1 1} = a_{j_2j_2 1}$. In the vicinity of $y_{j_1}^* = a_{j_1j_1 1} = a_{j_2j_2 1}$, we have

$$\Delta\dot{y}_{j_1} = (\alpha_{j_1j_1}a_{j_1j_10} + \alpha_{j_1j_2}a_{j_2j_20})(\Delta y_{j_1})^2 = G^{(1)}_{j_1j_10}(\Delta y_{j_1})^2,$$

where

$$G^{(1)}_{j_1j_10} = \alpha_{j_1j_1}a_{j_1j_10} + \alpha_{j_1j_2}a_{j_2j_20}.$$

The inflection flows at $y^*_{j_1} = a_{j_1j_11} = a_{j_2j_21}$ are upper-saddle and lower-saddle, respectively.

Therefore, the infinite-equilibrium of $y^*_{j_1} = a_{j_1j_11} = a_{j_2j_21}$ has the following properties as in Eqs. (4.92)–(4.95).

- *For* $a_{j_1j_10}/a_{j_2j_20} > 0$ *and* $G^{(1)}_{j_1j_10} > 0$,

$$(a_{j_1j_11}, a_{j_2j_21}) = \underbrace{(\text{II}, \text{US})}_{\text{increasing-inflection upper-saddle}}.$$

- *For* $a_{j_1j_10}/a_{j_2j_20} > 0$ *and* $G^{(1)}_{j_1j_10} > 0$,

$$(a_{j_1j_11}, a_{j_2j_21}) = \underbrace{(\text{II}, \text{LS})}_{\text{increasing-inflection lower-saddle}}.$$

- *For* $a_{j_1j_10}/a_{j_2j_20} < 0$ *and* $G^{(1)}_{j_1j_10} > 0$,

$$(a_{j_1j_11}, a_{j_2j_21}) = \underbrace{(\text{DI}, \text{US})}_{\text{decreasing-inflection upper-saddle}}.$$

- *For* $a_{j_1j_10}/a_{j_2j_20} < 0$ *and* $G^{(1)}_{j_1j_10} > 0$,

$$(a_{j_1j_11}, a_{j_2j_21}) = \underbrace{(\text{DI}, \text{LS})}_{\text{decreasing-inflection lower-saddle}}.$$

The infinite-equilibriums are the switching bifurcations for the up-parabola and down-parabola flows in the x_{j_1} and x_{j_2}-directions.

From the above local analysis of such a dynamical system, this theorem is proved. ■

4.3 Illustrations

In this section, dynamical systems with two quadratic vector fields with a single bivariate function will be discussed as in Luo (2023).

4.3.1 Inflection and Parabola-Flows

Consider

$$\begin{aligned}\dot{x}_1 &= a_{110}(y_1 - a_{111})(y_1 - a_{112}),\\ \dot{x}_2 &= a_{220}(y_1 - a_{221})^2\end{aligned} \tag{4.96}$$

with

$$y_1 = \alpha_{11}x_1 + \alpha_{12}x_2. \tag{4.97}$$

There is a dynamical system in the y_1-direction as

$$\dot{y}_1 = \alpha_{11}a_{110}(y_1 - a_{111})(y_1 - a_{112}) + \alpha_{12}a_{220}(y_1 - a_{221})^2, \tag{4.98}$$

with a dynamical system in the $y_{j_1}^{\perp}$-direction as

$$\dot{y}_1^{\perp} = \beta_{11}a_{110}(y_1 - a_{111})(y_1 - a_{112}) + \beta_{12}a_{220}(y_1 - a_{221})^2, \tag{4.99}$$

where

$$\begin{aligned}y_1^{\perp} &= \beta_{11}x_1 + \beta_{12}x_2,\\ \beta_{11}\alpha_{11} &+ \beta_{12}\alpha_{12} = 0.\end{aligned} \tag{4.100}$$

For $a_{111} > a_{221}$ ($\alpha_{11} > 0$, $\alpha_{12} > 0$), the phase portraits for the two-dimensional systems on the x_1-direction with $y_1^* = a_{111}, a_{112}$ and on the x_2-direction with $y_1^* = a_{221}$ are presented in Fig. 4.1(a)-(d) for the (DP,pF) and (UP,pF)-flows of $y_1^* = a_{111}, a_{112}$ and (pF,II)-flow of $y_1^* = a_{221}$ ($a_{110} > 0$, $a_{220} > 0$), the (UP,pF) and (DP,pF)-flows of $y_1^* = a_{111}, a_{112}$ and (nF,DI)-flow of $y_1^* = a_{221}$ ($a_{110} < 0$, $a_{220} > 0$), the (UP,nF) and (DP,nF)-flows of $y_1^* = a_{111}, a_{112}$ and (pF,DI)-flow of $y_1^* = a_{221}$ ($a_{110} > 0$, $a_{220} < 0$), and the (DP,nF) and (UP,nF)-flows of $y_1^* = a_{111}, a_{112}$ and (nF,II)-flow of $y_1^* = a_{221}$ ($a_{110} < 0, a_{220} < 0$), respectively. The solid lines are for $y_1^* = a_{111}, a_{112}$ and $y_1^* = a_{221}$. The dashed lines for the equilibrium of $\dot{y}_1 = 0$ are depicted. The decreasing and increasing inflection flows bifurcation for up-parabola and down-parabola flows exist at $y_1^* = a_{221}$.

Similarly, for $a_{111} > a_{221}$ ($\alpha_{11} > 0$, $\alpha_{12} < 0$), the phase portraits for the two-dimensional systems on the x_1-direction with $y_1^* = a_{111}, a_{112}$ and on the x_2-direction with $y_1^* = a_{221}$ are presented in Fig. 4.2(a)-(d) for the (UP,pF) and (DP, pF)-flows of $y_1^* = a_{111}, a_{112}$ and (pF,II)-flow of $y_1^* = a_{221}$ ($a_{110} > 0$, $a_{220} > 0$), the (DP,pF) and (UP,pF)-flows of $y_1^* = a_{111}, a_{112}$ and (nF,DI)-flow of $y_1^* = a_{221}$ ($a_{110} < 0$, $a_{220} > 0$), the (DP,nF) and (UP,nF)-flows of $y_1^* = a_{111}, a_{112}$ and (pF, DI)-flow of $y_1^* = a_{221}$ ($a_{110} > 0$, $a_{220} < 0$), and the (UP,nF) and (DP,nF)-flows of

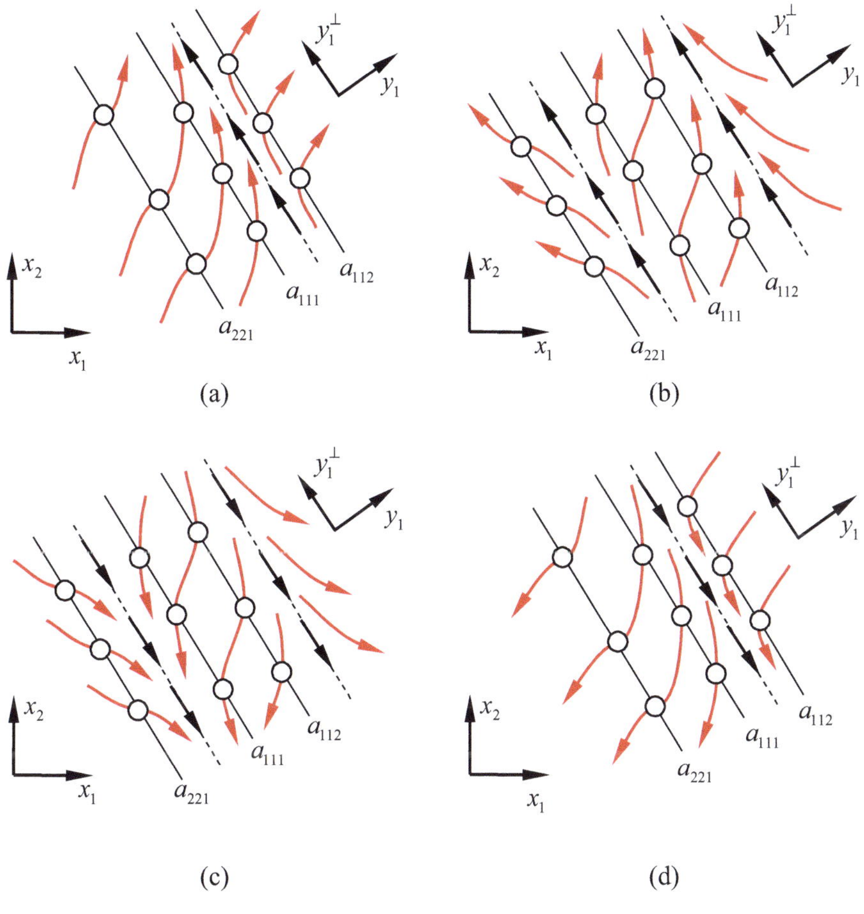

Fig. 4.1 Phase portraits ($a_{111} > a_{221}$ and $\alpha_{11} > 0$, $\alpha_{12} > 0$) for two-dimensional systems on the x_1-direction with $y_1^* = a_{111}, a_{112}$ and on the x_2-direction with $y_1^* = a_{221}$: (**a**) (DP,pF) and (UP,pF)-flows of $y_1^* = a_{111}, a_{112}$ and (pF,II)-flow of $y_1^* = a_{221}$ ($a_{110} > 0$, $a_{220} > 0$), (**b**) (UP,pF) and (DP,pF)-flows of $y_1^* = a_{111}, a_{112}$ and (nF,DI)-flow of $y_1^* = a_{221}$ ($a_{110} < 0$, $a_{220} > 0$), (**c**) (UP,nF) and (DP,nF)-flows of $y_1^* = a_{111}, a_{112}$ and (pF,DI)-flow of $y_1^* = a_{221}$ ($a_{110} > 0$, $a_{220} < 0$), (**d**) (DP,nF) and (UP,nF)-flows of $y_1^* = a_{111}, a_{112}$ and (nF,II)-flow of $y_1^* = a_{221}$ ($a_{110} < 0$, $a_{220} < 0$)

$y_1^* = a_{111}, a_{112}$ and (nF,II)-flow of $y_1^* = a_{221}$ ($a_{110} < 0$, $a_{220} < 0$). Only the parabola flows of $y_1^* = a_{111}, a_{112}$ are different from for ($\alpha_{11} > 0$, $\alpha_{12} > 0$).

The phase portraits ($a_{112} < a_{221}$) and) for the two-dimensional systems on the x_1-direction with $y_1^* = a_{111}, a_{112}$ and on the x_2-direction with $y_1^* = a_{221}$ are presented in Figs. 4.3 and 4.4 for ($\alpha_{11} > 0$, $\alpha_{12} > 0$) and ($\alpha_{11} > 0$, $\alpha_{12} < 0$), respectively.

The phase portraits ($a_{111} < a_{221} < a_{112}$) and) for the two-dimensional systems on the x_1-direction with $y_1^* = a_{111}, a_{112}$ and on the x_2-direction with $y_1^* = a_{221}$ are presented in Figs. 4.5 and 4.6 for ($\alpha_{11} > 0$, $\alpha_{12} > 0$) and ($\alpha_{11} > 0$, $\alpha_{12} < 0$), respectively.

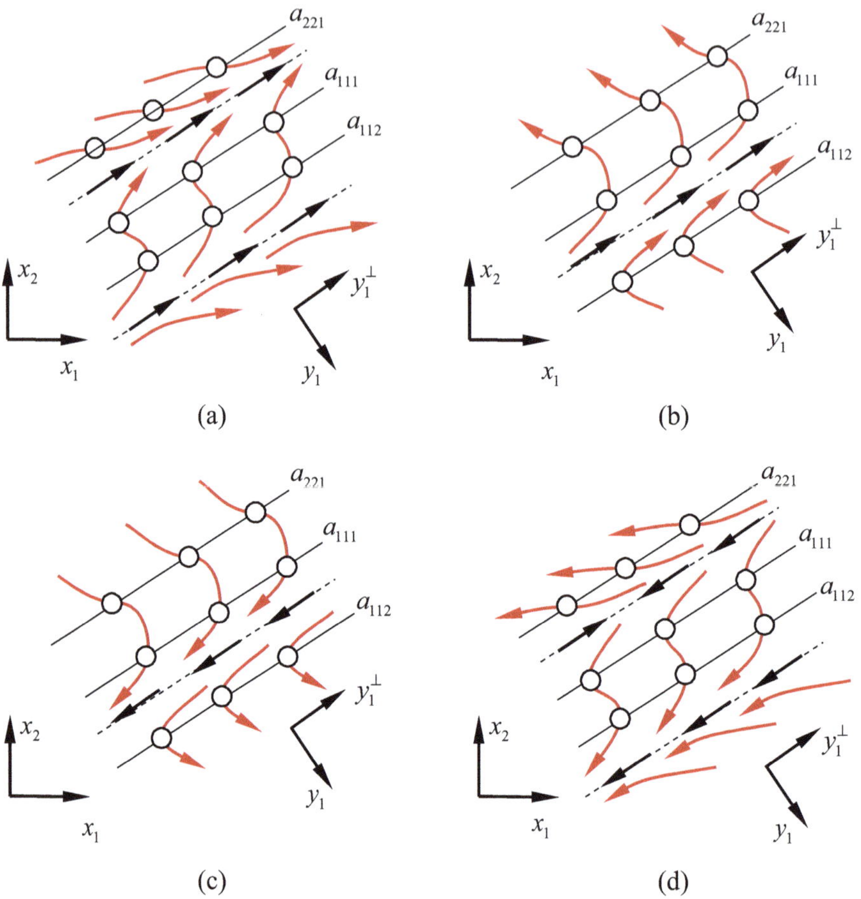

Fig. 4.2 Phase portraits ($a_{111} > a_{221}$ and $\alpha_{11} > 0$, $\alpha_{12} < 0$) for two-dimensional systems on the x_1-direction with $y_1^* = a_{111}, a_{112}$ and on the x_2-direction with $y_1^* = a_{221}$: (**a**) (UP,pF) and (DP,pF)-flows of $y_1^* = a_{111}, a_{112}$ and (pF,II)-flow of $y_1^* = a_{221}$ ($a_{110} > 0$, $a_{220} > 0$), (**b**) (DP,pF) and (UP,pF)-flows of $y_1^* = a_{111}, a_{112}$ and (nF,DI)-flow of $y_1^* = a_{221}$ ($a_{110} < 0$, $a_{220} > 0$), (**c**) (DP,nF) and (UP,nF)-flows of $y_1^* = a_{111}, a_{112}$ and (pF,DI)-flow of $y_1^* = a_{221}$ ($a_{110} > 0$, $a_{220} < 0$), (**d**) (UP,nF) and (DP,nF)-flows of $y_1^* = a_{111}, a_{112}$ and (nF,II)-flow of $y_1^* = a_{221}$ ($a_{110} < 0$, $a_{220} < 0$)

4.3.2 *Parabola-Source and Sink Infinite-Equilibriums*

Consider

$$\begin{aligned} \dot{x}_1 &= a_{110}(y_1 - a_{111})(y_1 - a_{112}), \\ \dot{x}_2 &= a_{220}(y_1 - a_{221})^2, \end{aligned} \tag{4.101}$$

where $a_{111} = a_{221} < a_{112}$

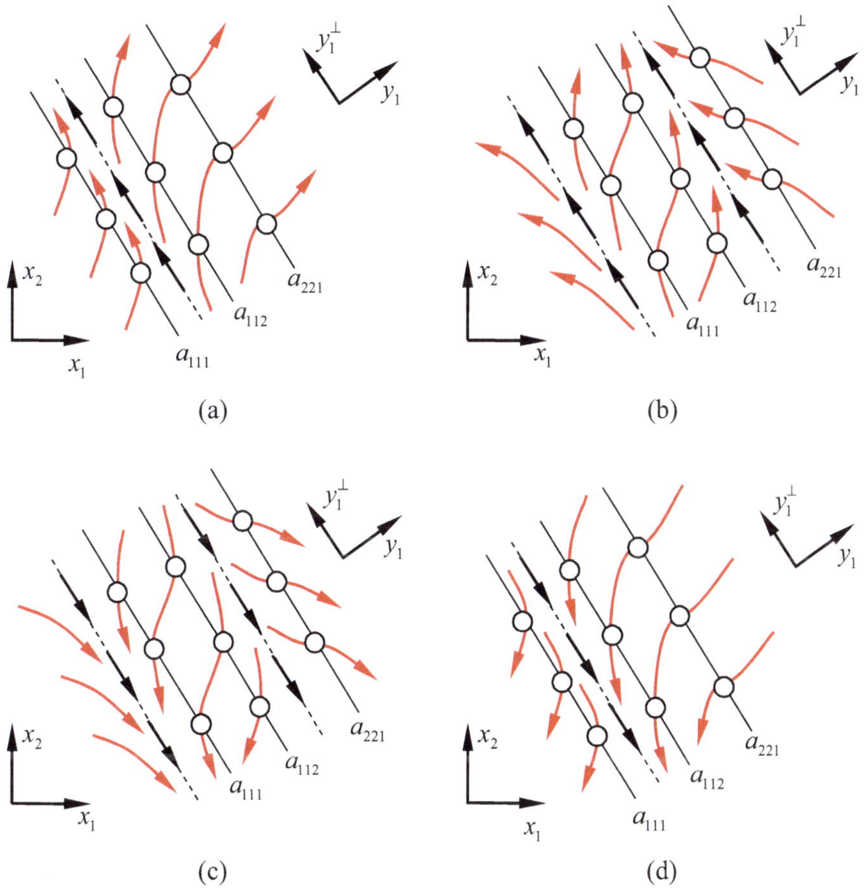

Fig. 4.3 Phase portraits ($a_{112} < a_{221}$ and $\alpha_{11} > 0$, $\alpha_{12} > 0$) for two-dimensional systems on the x_1-direction with $y_1^* = a_{111}, a_{112}$ and on the x_2-direction with $y_1^* = a_{221}$: (**a**) (DP,pF) and (UP,pF)-flows of $y_1^* = a_{111}, a_{112}$ and (pF,II)-flow of $y_1^* = a_{221}$ ($a_{110} > 0$, $a_{220} > 0$), (**b**) (UP,pF) and (DP,pF)-flows of $y_1^* = a_{111}, a_{112}$ and (nF,DI)-flow of $y_1^* = a_{221}$ ($a_{110} < 0$, $a_{220} > 0$), (**c**) (UP,nF) and (DP,nF)-flows of $y_1^* = a_{111}, a_{112}$ and (pF,DI)-flow of $y_1^* = a_{221}$ ($a_{110} > 0$, $a_{220} < 0$), (**d**) (DP,nF) and (UP,nF)-flows of $y_1^* = a_{111}, a_{112}$ and (nF,II)-flow of $y_1^* = a_{221}$ ($a_{110} < 0$, $a_{220} < 0$)

$$y_1 = \alpha_{11}x_1 + \alpha_{12}x_2. \tag{4.102}$$

There is a dynamical system in the y_1-direction as

$$\dot{y}_1 = (\alpha_{11}a_{110} + \alpha_{12}a_{220})(y_1 - a_{111})(y_1 - b_{111}), \tag{4.103}$$

with a dynamical system in the $y_{j_1}^\perp$-direction as

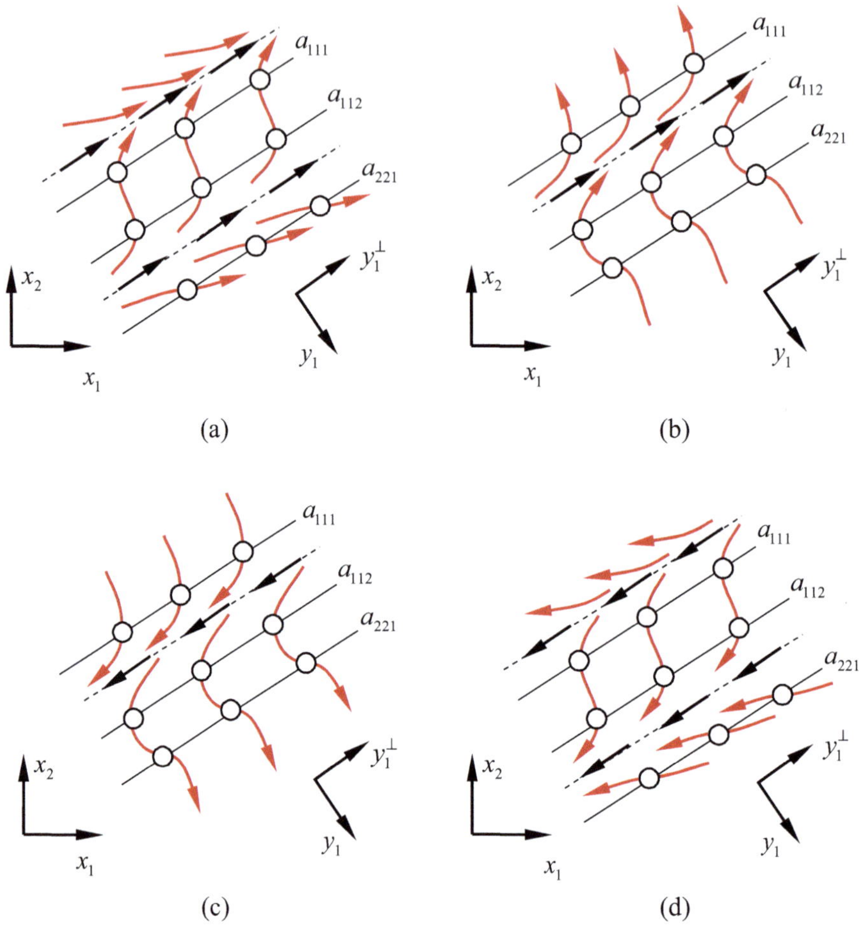

Fig. 4.4 Phase portraits ($a_{112} < a_{221}$ and $\alpha_{11} > 0$, $\alpha_{12} < 0$) for two-dimensional systems on the x_1-direction with $y_1^* = a_{111}, a_{112}$ and on the x_2-direction with $y_1^* = a_{221}$: (**a**) (UP,pF) and (DP,pF)-flows of $y_1^* = a_{111}, a_{112}$ and (pF,II)-flow of $y_1^* = a_{221}$ ($a_{110} > 0$, $a_{220} > 0$), (**b**) (DP,pF) and (UP,pF)-flows of $y_1^* = a_{111}, a_{112}$ and (nF,DI)-flow of $y_1^* = a_{221}$ ($a_{110} < 0$, $a_{220} > 0$), (**c**) (DP,nF) and (UP,nF)-flows of $y_1^* = a_{111}, a_{112}$ and (pF,DI)-flow of $y_1^* = a_{221}$ ($a_{110} > 0$, $a_{220} < 0$), (**d**) (UP,nF) and (DP,nF)-flows of $y_1^* = a_{111}, a_{112}$ and (nF,II)-flow of $y_1^* = a_{221}$ ($a_{110} < 0$, $a_{220} < 0$)

$$\dot{y}_1^\perp = (\beta_{11}a_{110} + \beta_{12}a_{220})(y_1 - a_{111})(y_1 - c_{111}), \tag{4.104}$$

where

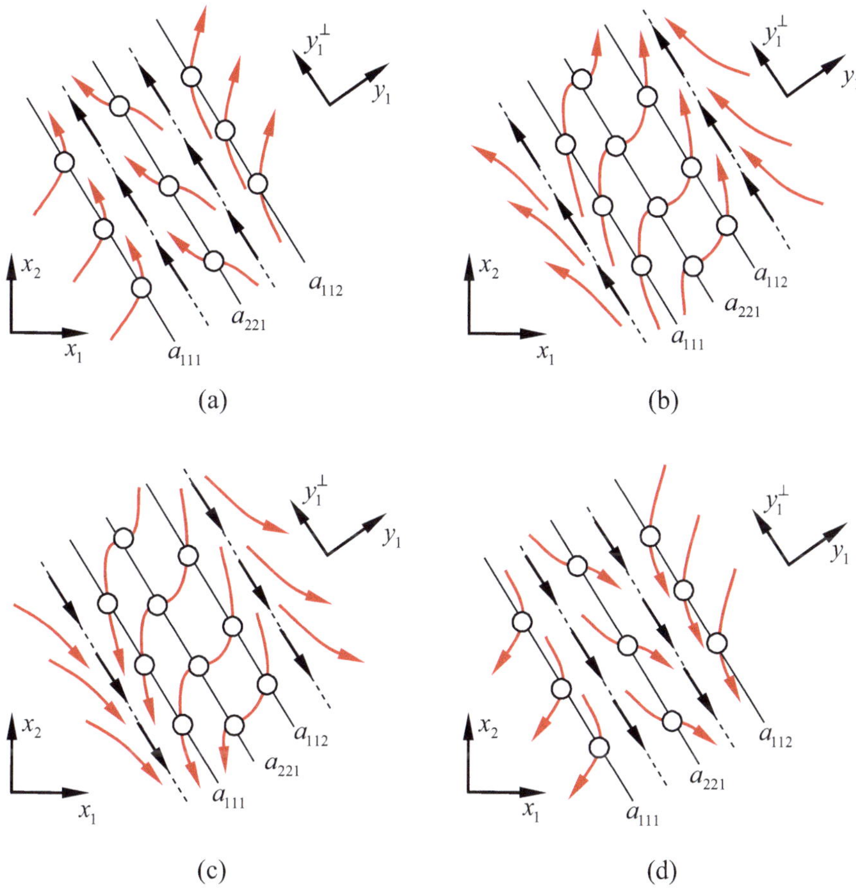

Fig. 4.5 Phase portraits ($a_{111} < a_{221} < a_{112}$ and $\alpha_{11} > 0$, $\alpha_{12} > 0$) for two-dimensional systems on the x_1-direction with $y_1^* = a_{111}, a_{112}$ and on the x_2-direction with $y_1^* = a_{221}$: (**a**) (DP,pF) and (UP,pF)-flows of $y_1^* = a_{111}, a_{112}$ and (pF,DI)-flow of $y_1^* = a_{221}$ ($a_{110} > 0$, $a_{220} > 0$), (**b**) (UP,pF) and (DP,pF)-flows of $y_1^* = a_{111}, a_{112}$ and (nF,II)-flow of $y_1^* = a_{221}$ ($a_{110} < 0$, $a_{220} < 0$), (**c**) (UP,nF) and (DP,nF)-flows of $y_1^* = a_{111}, a_{112}$ and (pF,II)-flow of $y_1^* = a_{221}$ ($a_{110} > 0$, $a_{220} < 0$), (**d**) (DP,nF) and (UP,nF)-flows of $y_1^* = a_{111}, a_{112}$ and (nF,DI)-flow of $y_1^* = a_{221}$ ($a_{110} < 0$, $a_{220} < 0$)

$$
\begin{aligned}
b_{111} &= \frac{\alpha_{11}a_{112} + \alpha_{12}a_{221}}{\alpha_{11}a_{110} + \alpha_{12}a_{220}}; \\
c_{111} &= \frac{\beta_{11}a_{112} + \beta_{12}a_{221}}{\beta_{11}a_{110} + \beta_{12}a_{220}}.
\end{aligned} \tag{4.105}
$$

For $a_{111} = a_{221}(\alpha_{11} > 0,\ \alpha_{12} > 0)$, the phase portraits for the two-dimensional system with the equilibrium of $y_1^* = a_{111}, a_{112}$ on the x_1-direction and with the equilibrium of $y_1^* = a_{221}$ on the x_2-direction are presented in Fig. 4.7. The thick solid line is for the infinite equilibrium. The thin solid line is for the one-dimensional flow, and the dash line is for the equilibrium on the y_1-direction. In Fig. 4.7a, the (SI,

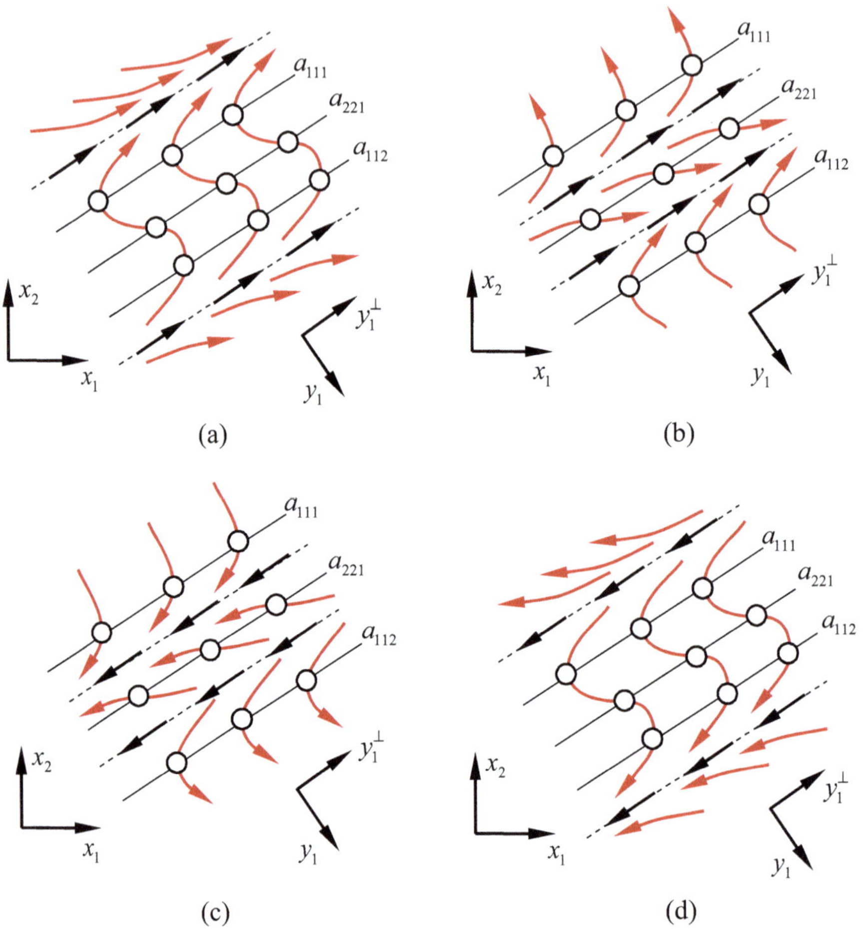

Fig. 4.6 Phase portraits ($a_{111} < a_{221} < a_{112}$ and $\alpha_{11} > 0$, $\alpha_{12} < 0$) for two-dimensional systems on the x_1-direction with $y_1^* = a_{111}, a_{112}$ and on the x_2-direction with $y_1^* = a_{221}$: (**a**) (UP,pF) and (DP,pF)-flows of $y_1^* = a_{111}, a_{112}$ and (nF,DI)-flow of $y_1^* = a_{221}$ ($a_{110} > 0$, $a_{220} > 0$), (**b**) (DP,pF) and (UP,pF)-flows of $y_1^* = a_{111}, a_{112}$ and (pF,II)-flow of $y_1^* = a_{221}$ ($a_{110} < 0$, $a_{220} > 0$), (**c**) (DP,nF) and (UP,nF)-flows of $y_1^* = a_{111}, a_{112}$ and (nF,II)-flow of $y_1^* = a_{221}$ ($a_{110} > 0$, $a_{220} < 0$), (**d**) (UP,pF) and (DP,nF)-flows of $y_1^* = a_{111}, a_{112}$ and (pF,DI)-flow of $y_1^* = a_{221}$ ($a_{110} < 0$, $a_{220} < 0$)

DP)-infinite-equilibrium of $y_1^* = a_{111} = a_{221}$ and the (UP,pF)-flow of $y_1^* = a_{112}$ are presented for ($a_{110} > 0$, $a_{220} > 0$). In Fig. 4.7b, the (SO,UP)-infinite-equilibrium of $y_1^* = a_{111} = a_{221}$ and the (DP,pF)-flow of $y_1^* = a_{112}$ are presented for ($a_{110} < 0$, $a_{220} > 0$). Similarly, the (SI,UP)-infinite-equilibrium of $y_1^* = a_{111} = a_{221}$ and the (DP,nF)-flow of $y_1^* = a_{112}$ for ($a_{110} > 0$, $a_{220} < 0$) are shown in Fig. 4.7c. The (SO, DP)-infinite-equilibrium of $y_1^* = a_{111} = a_{221}$ and the (UP,nF)-flow of $y_1^* = a_{112}$ ($a_{110} < 0$, $a_{220} < 0$) are presented in Fig. 4.7d. The equilibriums of $y_1^* = b_{111}$ on

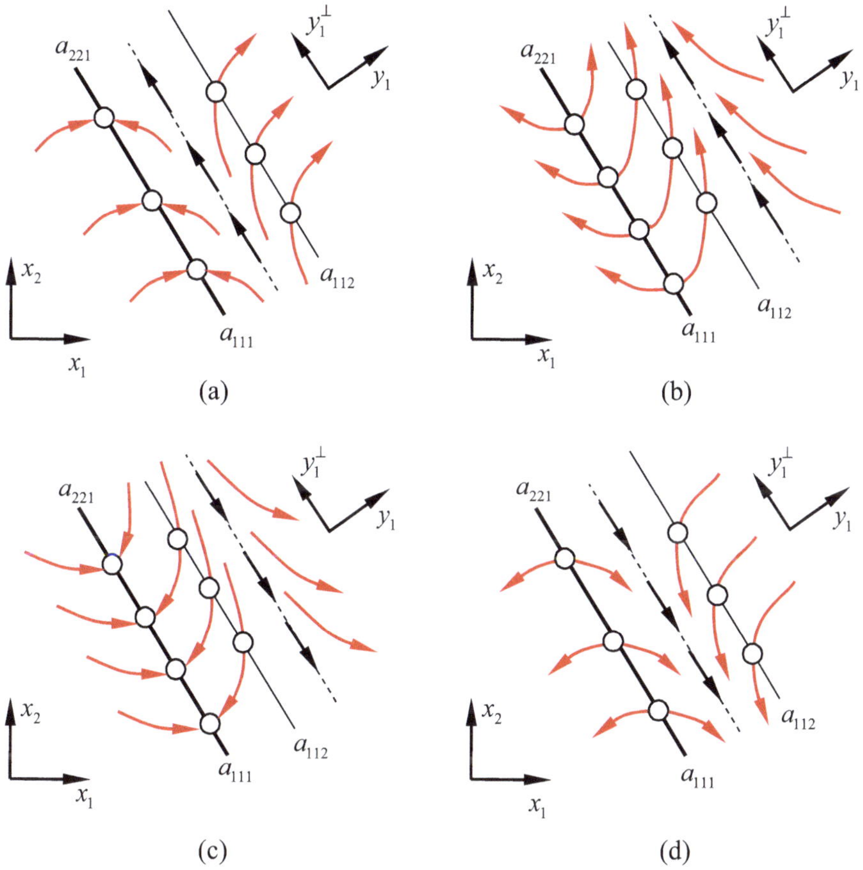

Fig. 4.7 Phase portraits ($a_{111} = a_{221}$ and $\alpha_{11} > 0$, $\alpha_{12} > 0$) for two-dimensional systems on the x_1-direction with $y_1^* = a_{111}, a_{112}$ and on the x_2-direction with $y_1^* = a_{221}$: (**a**) (SI,DP)-infinite-equilibrium of $y_1^* = a_{111} = a_{221}$ and (UP,pF)-flow of $y_1^* = a_{112}$ ($a_{110} > 0$, $a_{220} > 0$), (**b**) (SO,UP)-infinite-equilibrium of $y_1^* = a_{111} = a_{221}$ and (DP,pF)-flow of $y_1^* = a_{112}$ ($a_{110} < 0$, $a_{220} > 0$), (**c**) (SI,UP)-infinite-equilibrium of $y_1^* = a_{111} = a_{221}$ and (DP,nF)-flow of $y_1^* = a_{112}$ ($a_{110} > 0$, $a_{220} < 0$), (**d**) (SO, DP)-infinite-equilibrium of $y_1^* = a_{111} = a_{221}$ and (UP,nF)-flow of $y_1^* = a_{112}$ ($a_{110} < 0$, $a_{220} < 0$)

the y_1-direction are the one-dimensional source, sink, and source and sink flows are depicted in Fig. 4.7(a)-(d), respectively.

For $a_{111} = a_{221}$ ($\alpha_{11} > 0$, $\alpha_{12} < 0$), the phase portraits for the two-dimensional system with $y_1^* = a_{111}, a_{112}$ on the x_1-direction and with $y_1^* = a_{221}$ on the x_2-direction are presented in Fig. 4.8(a)-(d) for the (SI,DP)-infinite-equilibrium of $y_1^* = a_{111} = a_{221}$ and the (DP,pF)-flow of $y_1^* = a_{112}$ ($a_{110} > 0$, $a_{220} > 0$), the (SO,UP)-infinite-equilibrium of $y_1^* = a_{111} = a_{221}$ and the (UP,pF)-flow of $y_1^* = a_{112}$ ($a_{110} < 0$, $a_{220} > 0$), the (SI,UP)-infinite-equilibrium of $y_1^* = a_{111} = a_{221}$ and the (UP,nF)-flow of $y_1^* = a_{112}$ ($a_{110} > 0$, $a_{220} < 0$), the (SO,DP)-infinite-equilibrium of $y_1^* = a_{111} = a_{221}$ and the (DP,nF)-flow of $y_1^* = a_{112}$ ($a_{110} < 0$, $a_{220} < 0$), respectively. The

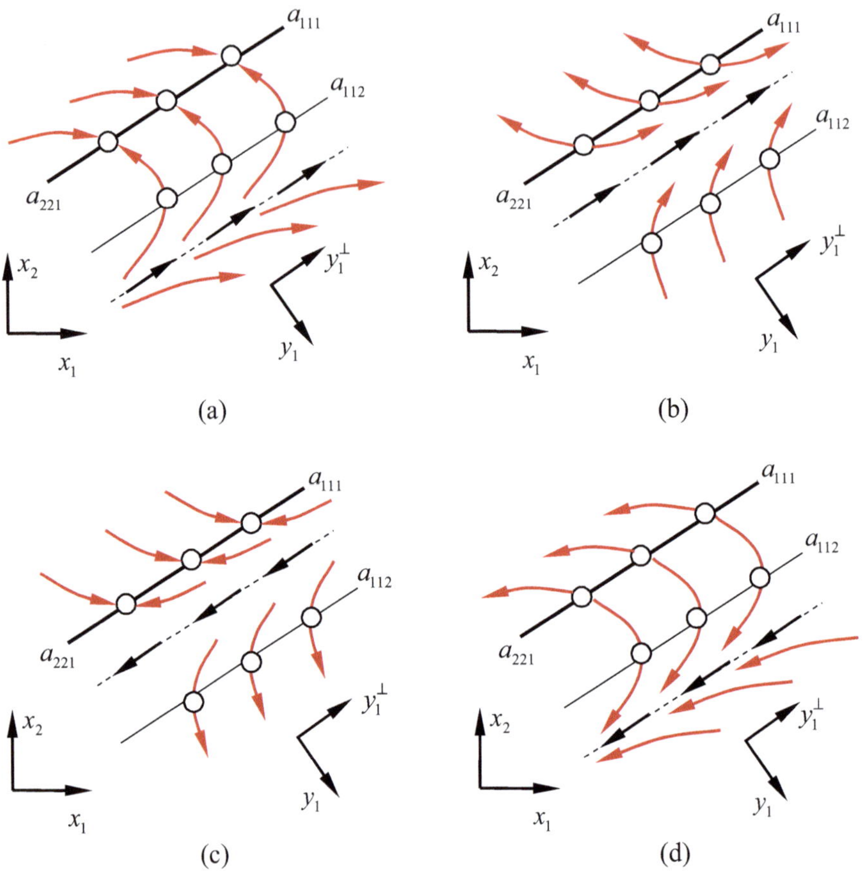

Fig. 4.8 Phase portraits ($a_{111} = a_{221}$ and $\alpha_{11} > 0$, $\alpha_{12} < 0$) for two-dimensional systems on the x_1-direction with $y_1^* = a_{111}, a_{112}$ and on the x_2-direction with $y_1^* = a_{221}$: (**a**) (SI,DP)-infinite-equilibrium of $y_1^* = a_{111} = a_{221}$ and (DP,pF)-flow of $y_1^* = a_{112}$ ($a_{110} > 0$, $a_{220} > 0$), (**b**) (SO,UP)-infinite-equilibrium of $y_1^* = a_{111} = a_{221}$ and (UP,pF)-flow of $y_1^* = a_{112}$ ($a_{110} < 0$, $a_{220} > 0$), (**c**) (SI,UP)-infinite-equilibrium of $y_1^* = a_{111} = a_{221}$ and (UP,nF)-flow of $y_1^* = a_{112}$ ($a_{110} > 0$, $a_{220} < 0$), (**d**) (SO, DP)-infinite-equilibrium of $y_1^* = a_{111} = a_{221}$ and (DP,nF)-flow of $y_1^* = a_{112}$ ($a_{110} < 0$, $a_{220} < 0$)

equilibriums of $y_1^* = b_{111}$ on the y_1-direction are the one-dimensional source, sink, and source and sink flows are the same as in Fig. 4.8.

For $a_{112} = a_{221}$ ($\alpha_{11} > 0$, $\alpha_{12} > 0$), the phase portraits for the two-dimensional system with $y_1^* = a_{111}, a_{112}$ on the x_1-direction and with $y_1^* = a_{221}$ on the x_2-direction are presented in Fig. 4.9. The (DP,pF)-flow of $y_1^* = a_{111}$ and the (SO,UP)-infinite-equilibrium of $y_1^* = a_{112} = a_{221}$ ($a_{110} > 0$, $a_{220} > 0$) are presented in Fig. 4.9a. The (UP,pF)-flow of $y_1^* = a_{111}$ and the (SI,DP)-infinite-equilibrium of $y_1^* = a_{112} = a_{221}$ are sketched in Fig. 4.9b for ($a_{110} < 0$, $a_{220} > 0$). In Fig. 4.9c, the (UP,nF)-flow of $y_1^* = a_{111}$ and (SO,DP)-infinite-equilibrium of $y_1^* = a_{112} = a_{221}$ ($a_{110} > 0$, $a_{220} < 0$) are presented. In Fig. 4.9d, the (DP,nF)-flow of $y_1^* = a_{111}$ and the (SI,UP)-infinite-

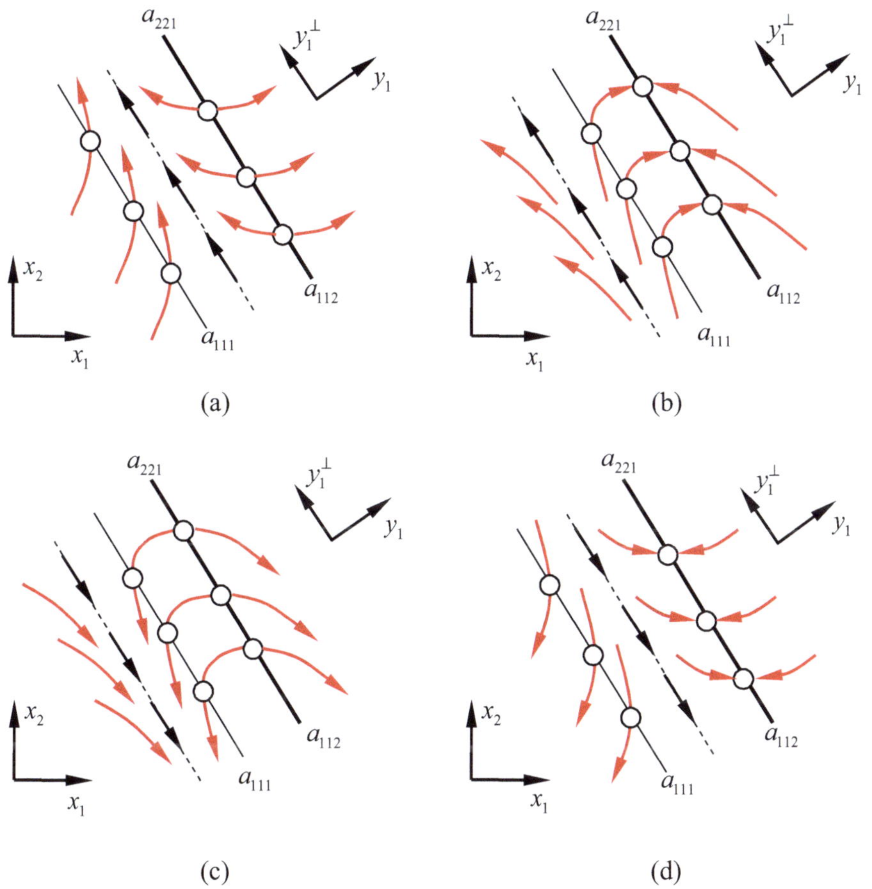

Fig. 4.9 Phase portraits ($a_{111} = a_{112}$ and $\alpha_{11} > 0$, $\alpha_{12} > 0$) for the two-dimensional system on the x_1-direction with $y_1^* = a_{111}, a_{112}$ and on the x_2-direction with $y_1^* = a_{221}$: (**a**) (DP,pF)-flow of $y_1^* = a_{111}$ and (SO,UP)-infinite-equilibrium of $y_1^* = a_{112} = a_{221}$ ($a_{110} > 0$, $a_{220} > 0$), (**b**) (UP,pF)-flow of $y_1^* = a_{111}$ and (SI,DP)-infinite-equilibrium of $y_1^* = a_{112} = a_{221}$ ($a_{110} < 0$, $a_{220} > 0$), (**c**) (UP,nF)-flow of $y_1^* = a_{111}$ and (SO,DP)-infinite-equilibrium of $y_1^* = a_{112} = a_{221}$ ($a_{110} > 0$, $a_{220} < 0$), (**d**) (DP,nF)-flow of $y_1^* = a_{111}$ and (SI,UP)-infinite-equilibrium of $y_1^* = a_{112} = a_{221}$ ($a_{110} < 0$, $a_{220} < 0$)

equilibrium of $y_1^* = a_{112} = a_{221}$ are shown in Fig. 4.9d for ($a_{110} < 0$, $a_{220} < 0$). The equilibriums of $y_1^* = b_{111}$ on the y_1-direction are the one-dimensional sink, source, and sink and source flows, as shown in Fig. 4.9(a)-(d), respectively.

Similarly, for $a_{112} = a_{221}(\alpha_{11} > 0,\ \alpha_{12} < 0)$, the phase portraits for a two-dimensional system on the x_1-direction with $y_1^* = a_{111}, a_{112}$ and on the x_2-direction with $y_1^* = a_{221}$ are presented in Fig. 4.10(a)-(d) for the (UP,pF)-flow of $y_1^* = a_{111}$ and the (SO,UP)-infinite-equilibrium of $y_1^* = a_{112} = a_{221}$ ($a_{110} > 0$, $a_{220} > 0$), the (DP,pF)-flow of $y_1^* = a_{111}$ and the (SI,DP)-infinite-equilibrium of $y_1^* = a_{112} = a_{221}$ ($a_{110} < 0$, $a_{220} > 0$), the (DP,nF)-flow of $y_1^* = a_{111}$ and the (SO,DP)-

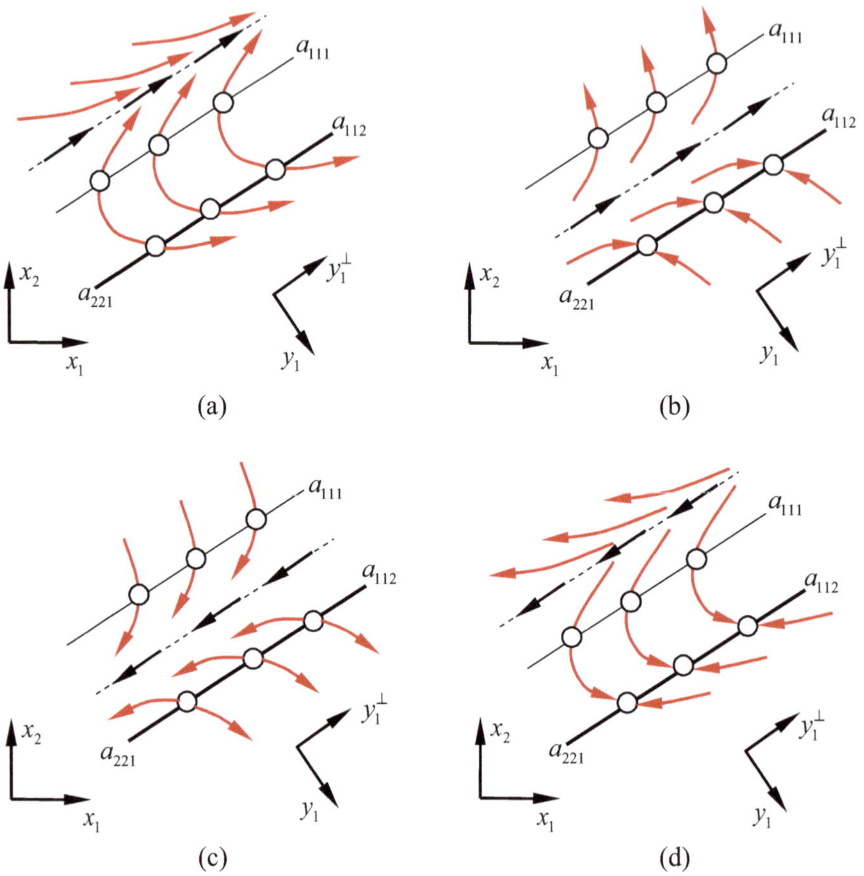

Fig. 4.10 Phase portraits ($a_{112} = a_{221}$ and $\alpha_{11} > 0, \alpha_{12} < 0$) for two-dimensional systems on the x_1-direction with $y_1^* = a_{111}, a_{112}$ and on the x_2-direction with $y_1^* = a_{221}$: (**a**) UP,pF)-flow of $y_1^* = a_{111}$ and (SO,UP)-infinite-equilibrium of $y_1^* = a_{112} = a_{221}$ ($a_{110} > 0$, $a_{220} > 0$), (**b**) (DP,pF)-flow of $y_1^* = a_{111}$ and (SI,DP)-infinite-equilibrium of $y_1^* = a_{112} = a_{221}$ ($a_{110} < 0$, $a_{220} > 0$), (**c**) (DP,nF)-flow of $y_1^* = a_{111}$ and (SO,DP)-infinite-equilibrium of $y_1^* = a_{112} = a_{221}$ ($a_{110} > 0$, $a_{220} < 0$), (**d**) (UP,nF)-flow of $y_1^* = a_{111}$ and (SI,UP)-infinite-equilibrium of $y_1^* = a_{112} = a_{221}$ ($a_{110} < 0$, $a_{220} < 0$)

infinite-equilibrium of $y_1^* = a_{112} = a_{221}$ ($a_{110} > 0$, $a_{220} < 0$), the (UP,nF)-flow of $y_1^* = a_{111}$ and the (SI,UP)-infinite-equilibrium of $y_1^* = a_{112} = a_{221}$ ($a_{110} < 0$, $a_{220} < 0$), respectively. The equilibriums of $y_1^* = b_{111}$ on the y_1-direction are the one-dimensional sink, source, and sink and source flows.

4.3.3 Double Inflection Flows

Consider

$$\begin{aligned}\dot{x}_1 &= a_{110}(y_1 - a_{111})^2,\\ \dot{x}_2 &= a_{220}(y_1 - a_{221})^2,\end{aligned} \tag{4.106}$$

with

$$y_1 = \alpha_{11}x_1 + \alpha_{12}x_2. \tag{4.107}$$

There is a dynamical system in the y_1-direction as

$$\dot{y}_1 = \alpha_{11}a_{110}(y_1 - a_{111})^2 + \alpha_{12}a_{220}(y_1 - a_{221})^2, \tag{4.108}$$

with a dynamical system in the $y_{j_1}^{\perp}$-direction as

$$\dot{y}_1^{\perp} = \beta_{11}a_{110}(y_1 - a_{111})^2 + \beta_{12}a_{220}(y_1 - a_{221})^2. \tag{4.109}$$

The equilibrium on the y_1-direction is given by

$$y_1^* = \frac{a_{111} + a_{221}\sqrt{-c_{11}}}{1 + \sqrt{-c_{11}}}, \frac{a_{111} - a_{221}\sqrt{-c_{11}}}{1 - \sqrt{-c_{11}}}, \tag{4.110}$$

where

$$c_{11} = \frac{\alpha_{12}a_{220}}{\alpha_{11}a_{110}} < 0. \tag{4.111}$$

For $a_{111} < a_{221}(\alpha_{11} > 0,\ \alpha_{12} > 0)$, the phase portraits for two-dimensional systems with the equilibrium of $y_1^* = a_{111}$ on the x_1-direction and the equilibrium of $y_1^* = a_{221}$ on the x_2-direction are presented in Fig. 4.11(a)-(d) for the (II,pF)-flow and (pF,II)-flow ($a_{110} > 0$, $a_{220} > 0$), the (DI,pF)-flow and (nF,DI)-flow ($a_{110} < 0$, $a_{220} > 0$), the (DI,nF)-flow and (pF,DI)-flow ($a_{110} > 0$, $a_{220} < 0$), the (II,nF)-flow and (nF,II)-flow ($a_{110} < 0$, $a_{220} < 0$), respectively. If $c_{11} < 0$, two equilibriums of y_1^* with flows on the $y_{j_1}^{\perp}$-direction are observed in Fig. 4.11b, c.

For $a_{111} < 0$ ($\alpha_{11} > 0$, $\alpha_{12} < 0$), the phase portraits for two-dimensional systems with the equilibrium of $y_1^* = a_{111}$ on the x_1-direction and the equilibrium of $y_1^* = a_{221}$ on the x_2-direction are presented in Fig. 4.12(a)-(d) for the (II,pF)-flow and (pF,II)-flow ($a_{110} > 0$, $a_{220} > 0$), the (DI,pF)-flow and (nF,DI)-flow ($a_{110} < 0$, $a_{220} > 0$), the (DI,nF)-flow and (pF,DI)-flow ($a_{110} > 0$, $a_{220} < 0$), the (II,nF)-flow and (nF,II)-flow ($a_{110} < 0$, $a_{220} < 0$), respectively. For $c_{11} < 0$, two equilibriums of y_1^* with flows on the $y_{j_1}^{\perp}$-direction are observed in Fig. 4.12a, d.

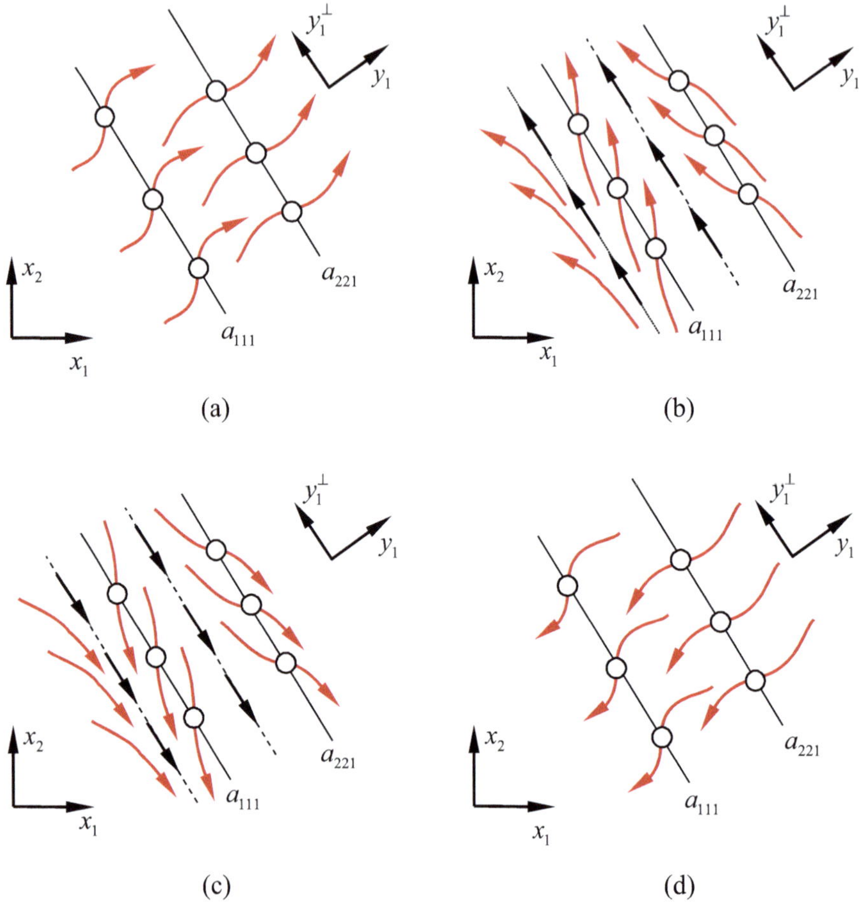

Fig. 4.11 Phase portraits ($a_{111} < a_{221}$ and $\alpha_{11} > 0, \alpha_{12} > 0$) for two-dimensional systems on the x_1-direction with $y_1^* = a_{111}$ and on the x_2-direction with $y_1^* = a_{221}$: (**a**) (II,pF)-flow of $y_1^* = a_{111}$ and (pF, II)-flow of $y_1^* = a_{112}$ ($a_{110} > 0, a_{220} > 0$), (**b**) (DI,pF)-flow of $y_1^* = a_{111}$ and (nF,DI)-flow of $y_1^* = a_{112}$ ($a_{110} < 0, a_{220} > 0$), (**c**) (DI,nF)-flow of $y_1^* = a_{111}$ and (pF,DI)-flow of $y_1^* = a_{112}$ ($a_{110} > 0, a_{220} < 0$), (**d**) (II,nF)-flow of $y_1^* = a_{111}$ and (nF,II)-flow of $y_1^* = a_{112}$ ($a_{110} < 0$, $a_{220} < 0$)

4.3.4 Inflection-Saddle Infinite-Equilibriums

Consider

$$\begin{aligned}\dot{x}_1 &= a_{110}(y_1 - a_{111})^2,\\ \dot{x}_2 &= a_{210}(y_1 - a_{221})^2.\end{aligned} \tag{4.112}$$

There is a dynamical system in the y_1-direction as

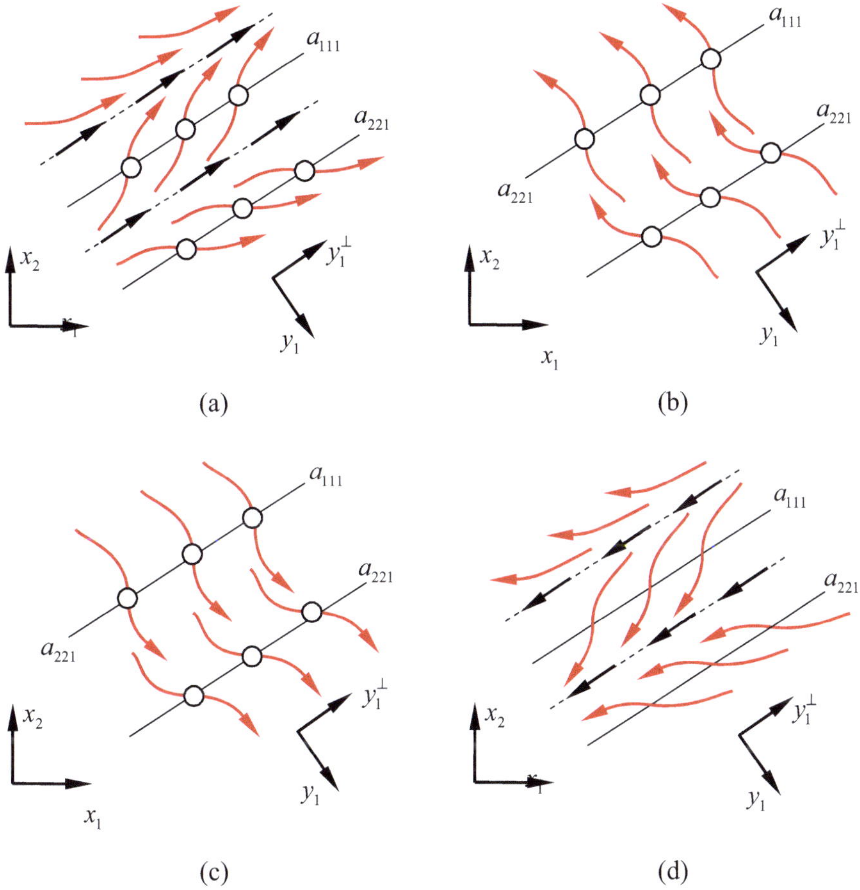

Fig. 4.12 Phase portraits ($a_{111} < a_{221}$ and $\alpha_{11} > 0$, $\alpha_{12} < 0$) for two-dimensional systems on the x_1-direction with $y_1^* = a_{111}$ and on the x_2-direction with $y_1^* = a_{221}$: (**a**) (II,pF)-flow of $y_1^* = a_{111}$ and (pF, II)-flow of $y_1^* = a_{112}$ ($a_{110} > 0$, $a_{220} > 0$), (**b**) (DI,pF)-flow of $y_1^* = a_{111}$ and (nF,DI)-flow of $y_1^* = a_{112}$ ($a_{110} < 0$, $a_{220} > 0$), (**c**) (DI,nF)-flow of $y_1^* = a_{111}$ and (pF,DI)-flow of $y_1^* = a_{112}$ ($a_{110} > 0$, $a_{220} < 0$), (**d**) (II,nF)-flow of $y_1^* = a_{111}$ and (nF,II)-flow of $y_1^* = a_{112}$ ($a_{110} < 0$, $a_{220} < 0$)

$$\dot{y}_1 = (\alpha_{11}a_{110} + \alpha_{12}a_{220})(y_1 - a_{111})^2, \tag{4.113}$$

with a dynamical system in the $y_{j_1}^{\perp}$-direction as

$$\dot{y}_1^{\perp} = (\beta_{11}a_{110} + \beta_{12}a_{220})(y_1 - a_{111})^2, \tag{4.114}$$

and

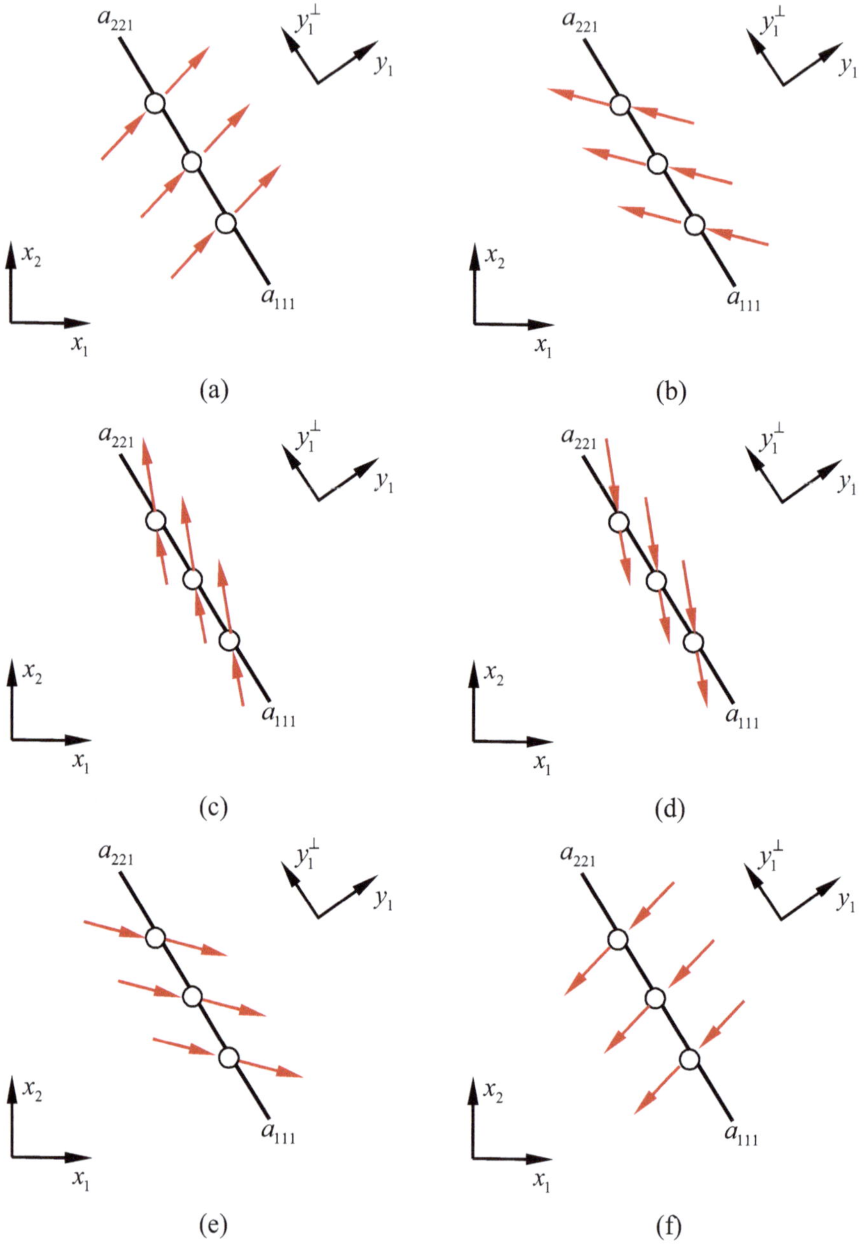

Fig. 4.13 Phase portraits ($a_{111} = a_{221}$ and $\alpha_{11} > 0, \alpha_{12} > 0$) for two-dimensional systems on the x_1-direction with $y_1^* = a_{111}$ and on the x_2-direction with $y_1^* = a_{221}$: (**a**) ($a_{110} > 0$, $a_{220} > 0$), (**b**, **c**) ($a_{110} < 0$, $a_{220} > 0$), (**d**, **e**) ($a_{110} > 0$, $a_{220} < 0$), (**f**) ($a_{110} < 0$, $a_{220} < 0$)

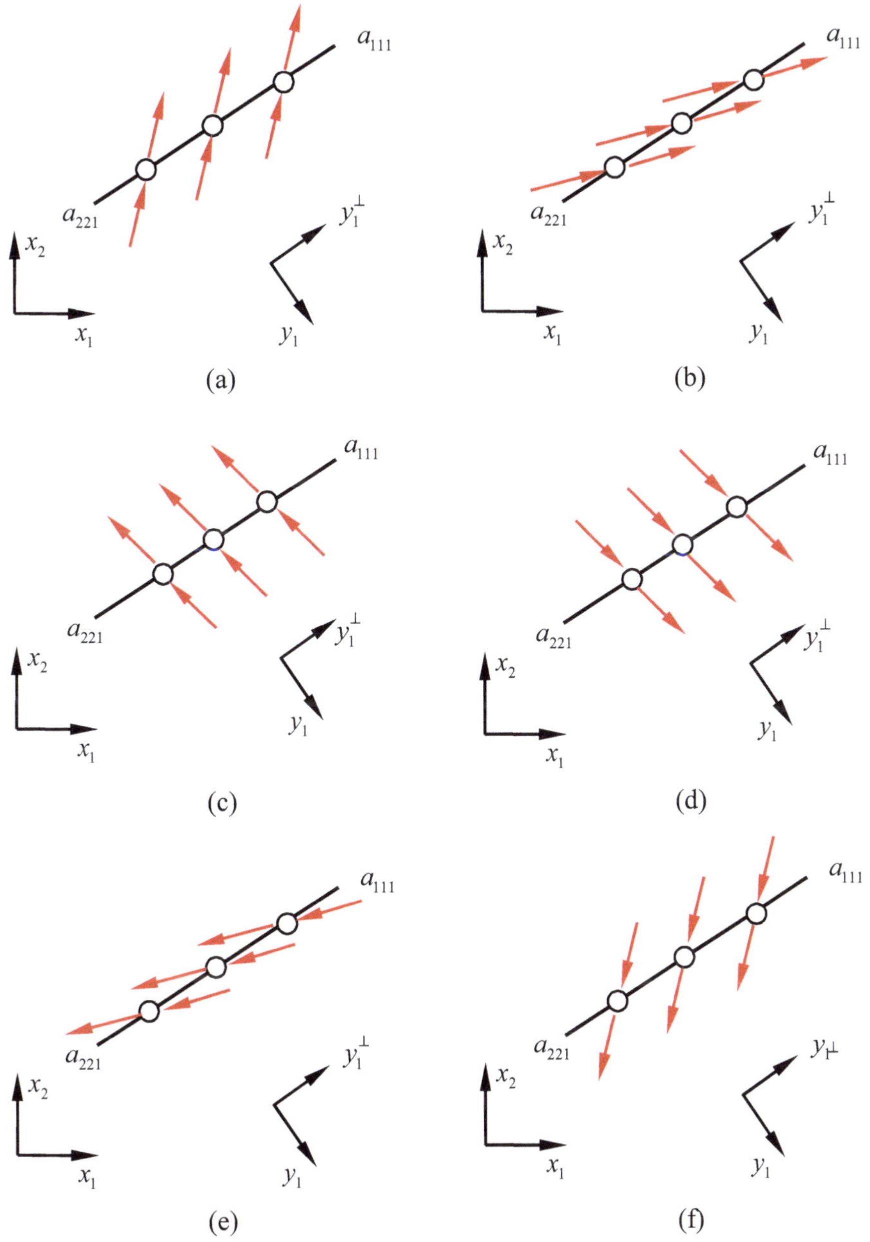

Fig. 4.14 Phase portraits ($a_{111} = a_{221}$ and $\alpha_{11} > 0$, $\alpha_{12} < 0$) for two-dimensional systems on the x_1-direction with $y_1^* = a_{111}$ and on the x_2-direction with $y_1^* = a_{221}$: (**a**, **b**) ($a_{110} > 0$, $a_{220} > 0$), (**c**) ($a_{110} < 0$, $a_{220} > 0$), (**d**) ($a_{110} > 0$, $a_{220} < 0$), (**e**, **f**) ($a_{110} < 0$, $a_{220} < 0$)

$$G_{110}^{(1)} = \alpha_{11}a_{110} + \alpha_{12}a_{220}. \quad (4.115)$$

For $a_{111} = a_{221}(\alpha_{11} > 0, \alpha_{12} > 0)$, the phase portraits for a two-dimensional system with the second-order infinite equilibrium of $y_1^* = a_{111} = a_{221}$ on the x_1-direction and on the x_2-direction are presented in Fig. 4.13(a)-(f) for the (II,US)-infinite-equilibrium with $G_{110}^{(1)} > 0$ $(a_{110} > 0, a_{220} > 0)$, the (DI,LS)-infinite-equilibrium with $G_{110}^{(1)} < 0$ and the (DI,US)-infinite-equilibrium with $G_{110}^{(1)} > 0(a_{110} < 0, a_{220} > 0$), the (DI,LS)-infinite-equilibrium with $G_{110}^{(1)} < 0$ and the (DI,US)-infinite-equilibrium with $G_{110}^{(1)} > 0$ $(a_{110} > 0, a_{220} < 0)$, and the (II,LS)-infinite equilibrium with $G_{110}^{(1)} < 0$ $(a_{110} < 0, a_{220} < 0)$, respectively.

For $a_{111} = a_{221}$ $(\alpha_{11} > 0, \alpha_{12} < 0)$, the phase portraits for a two-dimensional system with the second-order infinite equilibrium of $y_1^* = a_{111} = a_{221}$ on the x_1-direction and on the x_2-direction are presented in Fig. 4.14(a)-(f) for the (II, LS)-infinite-equilibrium with $G_{110}^{(1)} < 0$ and the (II,US)-infinite-equilibrium with $G_{110}^{(1)} > 0$ $(a_{110} > 0, a_{220} > 0)$, the (DI,LS)-infinite-equilibrium with $G_{110}^{(1)} < 0$ $(a_{110} < 0, a_{220} > 0$), the (DI,US)-infinite-equilibrium with $G_{110}^{(1)} > 0$ $(a_{110} > 0, a_{220} < 0)$, and the (II,LS)-infinite equilibrium with $G_{110}^{(1)} < 0$ and the (II,US)-infinite equilibrium with $G_{110}^{(1)} > 0$ $(a_{110} < 0, a_{220} < 0)$, respectively.

The infinite-equilibriums are inflection upper-saddles and lower-saddles.

Index

A. C. J. Luo, *Introduction to Infinite-Equilibriums in Dynamical Systems*,
https://doi.org/10.1007/978-3-031-89083-3

MIX
Papier aus verantwortungsvollen Quellen
Paper from responsible sources
FSC® C105338

If you have any concerns about our products,
you can contact us on
ProductSafety@springernature.com

In case Publisher is established outside the EU,
the EU authorized representative is:
Springer Nature Customer Service Center GmbH
Europaplatz 3, 69115 Heidelberg, Germany

Printed by Libri Plureos GmbH
in Hamburg, Germany